Biochemical Techniques and Instrumentation

THE AUTHORS

Mr F Parthiban is the Assistant Professor of Tamil Nadu Dr. J. Jayalalithaa Fisheries University, Nagapattinam. He has 10 years of fisheries industrial experience as managers in aquaculture, fish processing, and quality control at Saudi Arabia, Nigeria, and Tanzania and 4 years of experience in teaching, research and extension programmes. He has Recipient of CIFE, ICAR fellowship for Master of Fisheries Science in Post Harvest Technology at Central Institute of Fisheries Technology. He has successfully completed internal auditor and lead auditor courses for Food Safety Management System, ISO 22000:2005. He is the member of society of Fisheries Technologist and Professional Fisheries Graduates Forum in India. In addition, he has participated in many National and International Conferences, Symposia, Workshops, Seminars, etc., organized by fisheries development organizations. He has good expertise in aquatic food processing, aquatic food safety and quality. He has been handling courses for fisheries graduate on Fundamentals of Biochemistry, Biochemical Techniques and instrumentation, Microbiology of fish and fishery product, Fish Processing Technology. He has been conferred with "Best Teacher Award" by the Tamil Nadu Fisheries University for the year 2017. He has published research and extension articles in national and international journal.

Dr S Felix, Ph.D., Vice Chancellor, Tamil Nadu Dr. J. Jayalalithaa Fisheries University, Nagapattinam is having experience of more than 35 years in the area of fisheries and aquaculture in teaching, research, extension and administration. He has organized more than 20 Nos. of seminar/conference/symposium at National and International levels. Currently he is also the President (2018-19) of World Aquaculture Society, Asian Pacific Chapter and Chairman, ICAR's BSMA Committee (2018-19). He has more than 60 research papers published in National and International journals. He has authored more than 10 books and 25 manuals. He is specialized in the area of advanced systems in aquaculture and aquariculture such as raceway, biofloc technology, RAS, etc,. He has operated around 20 externally funded research projects.

डॉ. जे. के. जेना
उप महानिदेशक (मत्स्य विज्ञान)

Dr. J. K. Jena
Deputy Director General (Fisheries Science)

भारतीय कृषि अनुसंधान परिषद
कृषि अनुसंधान भवन-II, पूसा, नई दिल्ली 110 012
INDIAN COUNCIL OF AGRICULTURAL RESEARCH
KRISHI ANUSANDHAN BHAVAN-II, PUSA, NEW DELHI - 110 012

Ph. : 91-11-25846738 (O), Fax : 91-11-25841955
E-mail: ddgfs.icar@gov.in

Foreword

Biochemical studies rely on the availability of appropriate analytical techniques and its application to the advancement of knowledge of the nature and relationships between, biological molecules, especially proteins and nucleic acids, and cellular function. Biochemistry involves the study of the chemical processes that occur in living organisms with the ultimate aim of understanding the nature of life in molecular terms. The discipline of molecular biology overlaps with that of biochemistry and in many respects the aims of the two disciplines complement each other. Molecular biology is focussed on the molecular understanding of the processes of replication, transcription and translation of genetic material. Whereas biochemistry exploits the techniques and findings of molecular biology to advance understanding of cellular processes as cell signalling and apoptosis.

This book ***'Biochemical Techniques and Instrumentation'*** deals with the different biochemical techniques such as spectrophotometry and chromatography to gain knowledge on the biomolecules such as proteins, lipids, carbohydrates, nucleic acids and their functions. Molecular and immunological techniques such as ELISA, radioimmunoassay, blotting, PCR, cell culture, hybridoma and cloning protocols dealt in this are mainly focused to understand the diagnosis of diseases, malfunctions and disorders in order to generate corrective measures.

I congratulate the author, Drs. F. Parthiban and S. Felix for their efforts in bringing out this useful publication.

(J.K. Jena)

Message

It gives me immense pleasure to write the forward for this book that is primarily meant for the use of professionals in the discipline of fish biochemistry and molecular biology. It will be useful for the students, researchers and teachers of the life science. This book provides basic description of biochemical techniques, theoretical principles, instrumentation parts and their application in aquaculture and fisheries. I am sure that this book will help the readers, especially the students to gain basic understanding to develop skills in instrumentation for various analysis.

TNJFU has sophisticated centralized instrumentation laboratories to carry out of disease diagnosis and certification of various fish and fishery products. The university is also conducting various skill developments training program on instrumentation methods involved in fisheries science. I hope that this book will appeal to all students, teachers and researchers belonging to the disciplines of biochemistry and molecular biology in particular with aquaculture and fisheries.

S. Felix
Vice Chancellor
Tamil Nadu Dr. J. Jayalalithaa Fisheries University

Preface

Indian fish production through wild catch and aquaculture has steadily increased to approximately 11.41 million metric tonnes in the years 2016-17. During the year 2016-17, a sum of Rs. 37,870.90 crore has been earned as foreign exchange through the export of Rs.11.34 lakh tonnes of fish products. Tamil Nadu has exported Rs. 5,308 crore worth seafood products in this year. Aquaculture and seafood industries are challenged with various issues to ensure sustainable production and consumer safety. The methods for detection and quantification are getting essential now days. It is mandatory to enrich the students knowledge with basic and advanced techniques involved in various assays *viz.* HPLC, GC, PCR *etc.* Biochemical Techniques and Instrumentation is the course offered to the B. F. Sc students to get acquaint with various techniques applicable in the field of aquaculture and fisheries.

This book consists of information on the various biochemical techniques starting from spectroscopic, chromatographic, and centrifugation methods used in fish quality analysis along with specific applications. This book provides basic description of each technique, their principle, instrumentation and their applications in fisheries science. This shall help students to gain in-depth knowledge on the instrumentation techniques. This book shall be of use not only for the students but also for those employed as aquaculturists, fish quality control technologists, researchers, teachers and scientists to gain the basics of analytical and molecular techniques.

Authors

Contents

Chapter 1

Introduction

Biochemical studies rely on the availability of appropriate analytical techniques and its application to the advancement of knowledge of the nature and relationships between, biological molecules, especially proteins and nucleic acids, and cellular function. Biochemistry involves the study of the chemical processes that occur in living organisms with the ultimate aim of understanding the nature of life in molecular terms. The discipline of molecular biology overlaps with that of biochemistry and in many respects the aims of the two disciplines complement each other. Molecular biology is focussed on the molecular understanding of the processes of replication, transcription and translation of genetic material. Whereas biochemistry is exploits the techniques and findings of molecular biology to advance understanding of cellular processes as cell signalling and apoptosis. The result is that the two disciplines now have the opportunity to address issues such as:

- ☆ The structure and function of the total protein component of the cell (proteomics) and of all the small molecules in the cell (metabolomics);
- ☆ The mechanisms involved in the control of gene expression;
- ☆ The identification of genes associated with a wide range of diseases;
- ☆ The development of gene therapy strategies for the treatment of diseases;
- ☆ The engineering of cells, especially stem cells, to treat diseases;
- ☆ The understanding of the functioning of the immune system in order to develop strategies for the protection against invading pathogens;
- ☆ The development of our knowledge of the molecular biology of plants in order to engineer crop improvements, pathogen resistance and stress tolerance;

- ✰ The application of molecular biology techniques to the nature and treatment of bacterial, fungal and viral diseases.

This manual deals with the different biochemical techniques such as spectrophotometry and chromatography to gain knowledge on the biomolecules such as proteins, lipids, carbohydrates, nucleic acids and their functions. Molecular and immunological techniques such as ELISA, radioimmunoassay, blotting, PCR, cell culture, hybridoma and cloning protocols dealt in this manual are mainly focused to understand the diagnosis of diseases, malfunctions and disorders in order to generate corrective measures.

Chapter 2

Spectrophotometry

2.1. Theory of Spectrophotometry

2.1.1. Introduction

A spectrophotometer consists of two instruments, namely a spectrometer for producing light of any selected color (wavelength), and a photometer for measuring the intensity of light. The instruments are arranged so that liquid in a cuvette can be placed between the spectrometer beam and the photometer. The amount of light passing through the tube is measured by the photometer. The photometer delivers a voltage signal to a display device, normally a galvanometer. The signal changes as the amount of light absorbed by the liquid changes.

Spectroscopic techniques involve the interaction of light with matter. Spectroscopic techniques are used to quantify a chemical substance and to study their structure and reactions. This technique is unique as it does not degrade the molecules. They are the first analytical procedures used by is the biochemical scientists. If development of colour is linked to the concentration of a substance in solution, then that concentration can be measured by determining the extent of absorption of light at the appropriate wavelength. For example haemoglobin appears red because the hemoglobin absorbs blue and green light rays much more effectively than red. The degree of absorbance of blue or green light is proportional to the concentration of hemoglobin.

2.1.2. Molecular Orbital Theory

This theory is modern and more rational. This theory assume that in molecules, atomic orbitals lose their identity and the electrons in molecules are present in new orbitals called molecular orbitals. In chemistry, molecular orbital (MO) theory is a method for determining molecular structure in which electrons are not assigned to individual bonds between atoms, but are treated as moving under the influence of the nuclei in the whole molecule. The spatial and energetic properties of electrons within atoms are fixed by quantum mechanics to form orbitals that contain these electrons. While atomic orbitals contain electrons ascribed to a single atom, molecular orbitals, which surround a number of atoms in a molecule, contain valence electrons between atoms.

A brief outline of this theory is given below:

- In a molecule, electrons are present in new orbitals called **molecular orbitals.**
- Molecular orbitals are formed by combination of atomic orbitals of equal energies (in case of homonuclear molecules) or of comparable energies (incase of heteronuclear molecules).
- The number of molecular orbitals formed is equal to the number of atomic orbitals undergoing combination.
- Two atomic orbitals can combine to form two molecular orbitals. One of these two molecular orbitals one has a lower energy and the other has a higher energy. The molecular orbital with lower energy is called **bonding molecular orbital** and the other with higher energy is called **anti bonding molecular orbital.**
- The shapes of molecular orbitals depend upon the shapes of combining atomic orbitals.
- The bonding molecular orbitals are represented by σ (sigma), π (pi), δ (delta) and the anti-bonding molecular orbitals are represented by σ^*, π^*, δ^*
- The molecular orbitals are filled in the increasing order of their energies, starting with orbital of least energy. **(Aufbau principle).**
- A molecular orbital can accommodate only two electrons and these two electrons must have opposite spins. **(Paul's exclusion principle).**
- While filling molecular orbitals of equal energy, pairing of electrons does not take place until all such molecular orbitals are singly filled with electrons having parallel spins. **(Hund's rule).**

Types of Molecular Transitions

- Electronic transition – electron between two orbitals(UV-vis).hí = Energy Difference of Orbitals
- Vibrational transitions – vibrational states associated with bonds holding molecule. Involve simultaneous:
- Rotational transitions – changes in rotational states, about its centre of gravity.

Types of Electronic Transitions

- ✰ to σ* Alkanes
- ✰ to π* Carbonyl compounds
- ✰ to π* Alkenes, carbonyl compounds, alkynes, azocompounds η to σ * Oxygen, nitrogen, sulfur and halogen compounds η to π* Carbonyl compounds

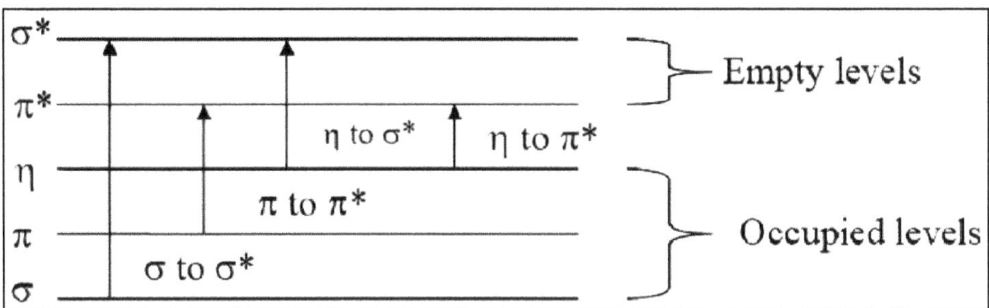

Atoms and molecules occupy distinct energy levels. They have unique energy configurations in terms of electron configurations, vibrational/rotational levels, *etc.* The Jablonski diagram gives the electronic and vibrational states of the molecule. Electrons in the atoms or molecules are distributed between several energy levels. Electrons principally reside in the lowest level or ground state (So). When energy is absorbed, it is promoted to a higher level or excited state (S1). The energy from electromagnetic radiation gives rise to this absorption spectrum. The molecule will also be in an excited vibrational and rotational state. The molecule will subsequent relax into the vibrational ground state of the first electronic excited state. The electron can then revert back to the electronic ground state. This will be accompanied by the emission of heat in case of non-fluorescent molecules.

The plot of absorption against wavelength is called absorption spectrum. Single atoms give line spectra, while molecules give band spectra due to different kinds of energy levels. Molecular spectra are molecule specific due to the unique vibrational states.

The energy diagram of molecules is more complex as they possess more than two atoms. The different atomic orbitals combine to yield molecular orbitals. When two atomic orbitals combine they form both a low energy bonding molecular orbital, and a high energy antibonding molecular orbital (*). The s and p atomic orbitals overlap in different ways. The head-on overlap of two s or two p atomic orbitals forms σ bond and the parallel overlap of two p atomic orbitals forms π bond. Single bonds are usually σ bonds, where as double bonds contain one σ bond and one π bond. Each of these molecular orbitals represents a different energy level.

Each atom in an organic molecule in its ground state or low energy state will contain bonding electrons in σ and π molecular orbitals, and outer, nonbonding unshared electrons 'n'. Molecular orbitals generally fall into one of the five different classes: s orbitals combine to the σ and the anti-bonding σ^* orbitals. Some p orbitals combine to the bonding π and the anti-bonding π^* orbitials. Other p orbitals combine to form non-bonding n orbitals. The population of bonding orbitals strengthens a chemical bond and the population of anti-bonding orbitals weakens a chemical bond.

2.1.3. Interaction of EMR with matter

Electromagnetic radiation (EM radiation or EMR) is the radiant energy released by certain electromagnetic processes. Visible light is one type of electromagnetic radiation; other familiar forms are invisible electromagnetic radiations, such as radio waves, infrared light and X rays. Classically, electromagnetic radiation consists of **electromagnetic waves**, which are synchronized oscillations of electric and magnetic fields that propagate at the speed of light through a vacuum. The oscillations of the two fields are perpendicular to each other and perpendicular to the direction of energy and wave propagation, forming a transverse wave. Electromagnetic (EM) waves can be characterized by either the frequency or wavelength of their oscillations to form the electromagnetic spectrum, Electromagnetic (EM) radiation is a form of energy that is all around us and takes many forms. Sunlight is also a form of EM energy, but visible light is only a small portion of the EM spectrum, which contains a broad range of electromagnetic wavelengths.

Electromagnetic waves are produced whenever charged particles are accelerated, and these waves can subsequently interact with any charged particles. EM waves carry energy, momentum and angular momentum away from their source particle and can impart those quantities to matter with which they interact. Quanta of EM waves are called photons, which are mass less, but they are still affected by gravity. Electromagnetic radiation is associated with those EM waves that are free to propagate themselves ("radiate") without the continuing influence of the moving charges that produced them, because they have achieved sufficient distance from those charges. Thus, EMR is sometimes referred to as the far field. In this language, the near field refers to EM fields near the charges and current that directly produced them, specifically, electromagnetic induction and electrostatic

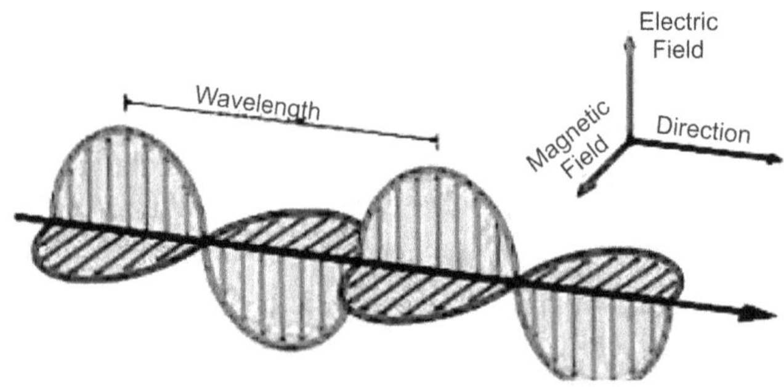

induction phenomena. The waves have certain characteristics, given as frequency, wavelength or energy.

A wavelength is the distance between two consecutive peaks of a wave. This distance is given in meters (m) or fractions thereof. Frequency is the number of waves that form in a given length of time. It is usually measured as the number of wave cycles per second, or hertz (Hz). A short wavelength means that the frequency will be higher because one cycle can pass in a shorter amount of time. Similarly, a longer wavelength has a lower frequency because each cycle takes longer to complete.

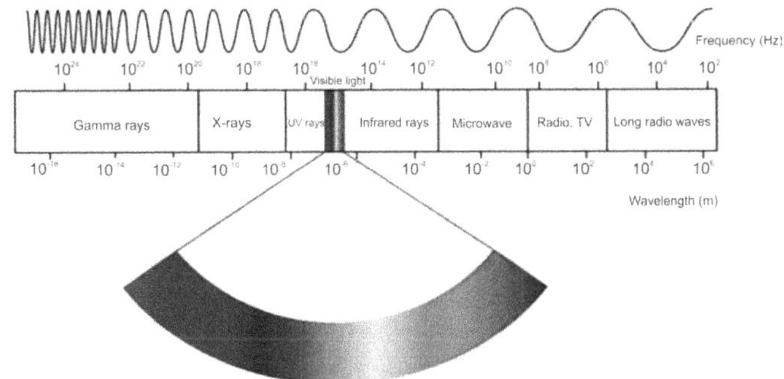

The EM spectrum is generally divided into seven regions, in order of decreasing wavelength and increasing energy and frequency. The common designations are: radio waves, microwaves, infrared (IR), visible light, ultraviolet (UV), X-rays and gamma rays. Typically, lower-energy radiation, such as radio waves, is expressed as frequency; microwaves, infrared, visible and UV light are usually expressed as wavelength; and higher-energy radiation, such as X-rays and gamma rays, is expressed in terms of energy per photon.

Radio Waves

Radio waves are at the lowest range of the EM spectrum, with frequencies of up to about 30 billion hertz, or 30 gigahertz (GHz), and wavelengths greater than about 10 millimetres (0.4 inches). Radio is used primarily for communications including voice, data and entertainment media.

Microwaves

Microwaves fall in the range of the EM spectrum between radio and IR. They have frequencies from about 3 GHz up to about 30 trillion hertz, or 30 terahertz (THz), and wavelengths of about 10 mm (0.4 inches) to 100 micrometers (μm), or 0.004 inches. Microwaves are used for high-bandwidth communications, radar and as a heat source for microwave ovens and industrial applications.

Infrared

Infrared is in the range of the EM spectrum between microwaves and visible light. IR has frequencies from about 30 THz up to about 400 THz and wavelengths of about 100 μm (0.004 inches) to 740 nanometers (nm), or 0.00003 inches. IR light is invisible to human eyes, but we can feel it as heat if the intensity is sufficient.

Visible Light

Visible light is found in the middle of the EM spectrum, between IR and UV. It has frequencies of about 400 THz to 800 THz and wavelengths of about 740 nm (0.00003 inches) to 380 nm (.000015 inches). More generally, visible light is defined as the wavelengths that are visible to most human eyes.

Ultraviolet

Ultraviolet light is in the range of the EM spectrum between visible light and X-rays. It has frequencies of about 8 × 1014 to 3 × 1016 Hz and wavelengths of about 380 nm (.000015 inches) to about 10 nm (0.0000004 inches). UV light is a component of sunlight; however, it is invisible to the human eye. It has numerous medical and industrial applications, but it can damage living tissue.

X-rays

X-rays are roughly classified into two types: soft X-rays and hard X-rays. Soft X-rays comprise the range of the EM spectrum between UV and gamma rays. Soft X-rays have frequencies of about 3 × 1016 to about 1018 Hz and wavelengths of about 10 nm (4×10^{-7} inches) to about 100 picometers (pm), or 4×10^{-8} inches. Hard X-rays occupy the same region of the EM spectrum as gamma rays. The only difference between them is their source: X-rays are produced by accelerating electrons, while gamma rays are produced by atomic nuclei.

Gamma-rays

Gamma-rays are in the range of the spectrum above soft X-rays. Gamma-rays have frequencies greater than about 1018 Hz and wavelengths of less than 100 pm (4×10^{-9} inches). Gamma radiation causes damage to living tissue, which makes it useful for killing cancer cells when applied in carefully measured doses to small regions. Uncontrolled exposure, though, is extremely dangerous to humans.

Light is electromagnetic radiation (EMR) that exhibit discrete packets of energy called photons. Electromagnetic spectrum consists of wavelength from the order of meters *i.e.* radio waves to less than 1 A° *i.e.* gamma rays. The wavelength of visible light is in a narrow range of between 4000 A° to 8000 A° (400 - 800nm). The transitions that occur due to the interaction of electromagnetic radiation (photons) with matter is a quantum phenomenon. The interaction is dependent upon the properties of the radiation and the appropriate structural components of the molecules. The photons interact with matter by transferring its energy.

Interaction of Electromagnetic Radiation and Matter

It is well known that all matter is comprised of atoms. But sub-atomically, matter is made up of mostly empty space. For example, consider the hydrogen atom with its one proton, one neutron, and one electron. The diameter of a single proton has been measured to be about 10-15 meters. The diameter of a single hydrogen atom has been determined to be 10-10 meters, therefore the ratio of the size of a hydrogen atom to the size of the proton is 100,000:1. Consider this in terms of something more easily pictured in your mind. If the nucleus of the atom could be enlarged to the size of a softball (about 10 cm), its electron would be approximately 10 kilometres away. Therefore, when electromagnetic waves pass through a material, they are primarily moving through free space, but may have a chance encounter with the nucleus or an electron of an atom.

Because the encounters of photons with atom particles are by chance, a given photon has a finite probability of passing completely through the medium it is traversing. The probability that a photon will pass completely through a medium depends on numerous factors including the photon's energy and the medium's composition and thickness. The more densely packed a medium's atoms, the more likely the photon will encounter an atomic particle. In other words, the more subatomic particles in a material (higher Z number), the greater the likelihood that interactions will occur Similarly, the more material a photon must cross through, the more likely the chance of an encounter.

When a photon does encounter an atomic particle, it transfers energy to the particle. The energy may be reemitted back the way it came (reflected), scattered in a different direction or transmitted forward into the material. Let us first consider the interaction of visible light. Reflection and transmission of light waves occur because the light waves transfer energy to the electrons of the material and cause them to vibrate. If the material is transparent, then the vibrations of the electrons are passed on to neighboring atoms through the bulk of the material and reemitted on the opposite side of the object. If the material is opaque, then the vibrations of the electrons are not passed from atom to atom through the bulk of the material, but rather the electrons vibrate for short periods of time and then reemit the energy as a reflected light wave. The light may be reemitted from the surface of the material at a different wavelength, thus changing its color.

X-Rays and Gamma Rays

X-rays and gamma rays also transfer their energy to matter though chance encounters with electrons and atomic nuclei. However, X-rays and gamma rays have enough energy to do more than just make the electrons vibrate. When these high energy rays encounter an atom, the result is an ejection of energetic electrons from the atom or the excitation of electrons. The term "excitation" is used to describe an interaction where electrons acquire energy from a passing charged particle but are not removed completely from their atom. Excited electrons may subsequently emit energy in the form of X-rays during the process of returning to a lower energy state.

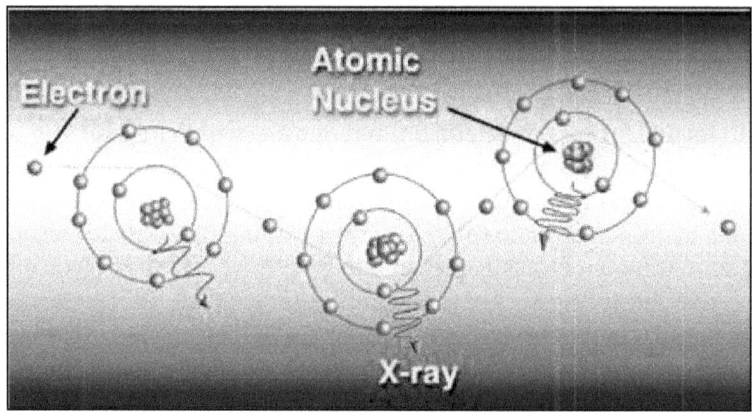

Each of the excited or liberated electrons goes on to transfer its energy to matter through thousands of events involving interactions between charged particles. With each interaction, the energy may be directed in a different direction. The higher the energy of a photon, the more likely the energy will continue traveling in the same direction. As the radiation moves from point to point in matter, it loses its energy through various interactions with the atoms it encounters. If the radiation has enough energy, it may eventually make it through the material.

2.1.4. UV-Vis Spectroscopy

Ultraviolet–visible spectroscopy or ultraviolet-visible spectrophotometry (UV-Vis or UV/Vis) refers to absorption spectroscopy or reflectance spectroscopy in the ultraviolet-visible spectral region. This means it uses light in the visible and adjacent (near-UV and near-infrared [NIR]) ranges. The absorption or reflectance in the visible range directly affects the perceived color of the chemicals involved. In this region of the electromagnetic spectrum, molecules undergo electronic transitions. This technique is complementary to fluorescence spectroscopy, in that fluorescence deals with transitions from the excited state to the ground state, while absorption measures transitions from the ground state to the excited state.

UV-Vis spectroscopy is based on the selective absorption of electromagnetic radiation in the wavelength range of 180-800 nm. Electronic transitions in molecules are classified according to the participating molecular orbitals. There are four possible transitions: $n \rightarrow \pi^*$, $\pi \rightarrow \pi^*$, $n \rightarrow \sigma^*$, $\sigma \rightarrow \sigma^*$. Transitions among these orbitals are by the radiation having the same energy as that of the energy difference between the specific orbitals. The UV-Vis radiation has sufficient energy to cause only two transitions between $n \rightarrow \pi^*$ and $\pi \rightarrow \pi^*$. The other transitions, $n \rightarrow \sigma^*$ and $\sigma \rightarrow \sigma^*$ requires higher energies. The molecular structures such as bonds and functional groups responsible for interaction with electromagnetic radiation are called **chromophores.**

In proteins, there are three types of chromophores: peptide bond, certain aminoacid side chains and certain prosthetic groups/coenzymes. Electronic transition of the peptide bond occur in the far UV with an intense peak at 190 nm and weak one at 210-220 nm due to $\pi \rightarrow \pi^*$ and $n \rightarrow \pi^*$ transitions. The aromatic aminoacids such as, phenylalanine, tyrosine and tryptophan have their absorption maxima at 257, 274 and 280 nm, respectively. The proteins that contain prosthetic groups such as haem, falvin, caroteniod have strong absorption bands in the UV-Vis region. The absorption of UV light by nucleic acids is due to $n \rightarrow \pi^*$ and $\pi \rightarrow \pi^*$ transitions of the purine and pyrimidine bases that occur between 260 nm and 275 nm.

2.1.5. Basic Laws of Light Absorption

The Beer-Lambert law relates the attenuation of light to the properties of the material through which the light is travelling. This page takes a brief look at the Beer-Lambert Law and explains the use of the terms absorbance and molar absorptivity relating to UV-visible absorption spectrometry. For each wavelength of light passing through the spectrometer, the intensity of the light passing through the reference cell is measured. This is usually referred to as Io - that's I for Intensity. The intensity of the light passing through the sample cell is also measured for that wavelength - given the symbol, I. If I is less than Io, then the sample has absorbed some of the light (neglecting reflection of light off the cuvette surface). A simple bit of math is then done in the computer to convert this into something called the absorbance of the sample - given the symbol, A.

The absorbance of a transition depends on two external assumptions.

The absorbance is directly proportional to the concentration (c) of the solution of the sample used in the experiment.

The absorbance is directly proportional to the length of the light path (l), which is equal to the width of the cuvette.

Deriving the Beer-Lambert Law

Assumption one relates the absorbance to concentration and can be expressed as

$$A \alpha c \quad (1)$$

The absorbance (A) is defined via the incident intensity Io and transmitted intensity I by

$$A=\log10(Io/I) \quad (2)$$

Assumption two can be expressed as

$$A \alpha l \quad (3)$$

Combining Equations 1 and 3:

$$A \alpha cl \quad (4)$$

This proportionality can be converted into equality by including proportionality constant.

$A = \varepsilon cl$ (5)

This formula is the common form of the Beer-Lambert Law, although it can be also written in terms of intensities:

$A = \log 10 (Io/I) = \varepsilon lc$ (6)

2.2. UV-V is Spectrophotometry and its Instrumentation

2.2.1. UV-V is Spectrophotometer

Ultraviolet-visible spectroscopy or **ultraviolet-visible spectrophotometry** (UV-Vis or UV/Vis) refers to absorption spectroscopy or reflectance spectroscopy in the ultraviolet-visible spectral region. Absorption spectroscopy refers to spectroscopic techniques that measure the absorption of radiation. Ultraviolet (UV) light is electromagnetic radiation with a wavelength shorter than that of visible light, but longer than X-rays, in the range 10 nm to 400 nm. The visible spectrum is the portion of the electromagnetic spectrum that is visible to (and can be detected by) the human eye, in the range of 390 to 750 nm. This means that UV spectrophotometry uses light in the visible and nearby (near-UV and near-infrared (NIR)) ranges.

The absorption or reflectance in the visible range directly affects the perceived color of the chemicals involved. The color of chemicals is a physical property of chemicals that in most cases comes from the excitation of electrons due to absorption of energy performed by the chemical. Excitation is an elevation in energy level above an arbitrary baseline energy state. In this region of the electromagnetic spectrum, molecules undergo electronic transitions. The electromagnetic spectrum is the range of all possible frequencies of electromagnetic radiation. Molecular electronic transitions take place when electrons in a molecule are excited from one energy level to a higher energy level.

What does the UV Spectrometer Measure?

The instrument used in ultraviolet-visible spectroscopy is called a **UV/V** is spectrophotometer. It measures the intensity of light passing through a sample (I), and compares it to the intensity of light before it passes through the sample (Io). The ratio I/Iois called the "transmittance", and is usually expressed as a percentage (per cent T). The absorbance, A, is based on the transmittance:

A = – log (T per cent/100 per cent)

The UV-visible spectrophotometer can also be configured to measure reflectance. In this case, the spectrophotometer measures the intensity of light reflected from a sample (I), and compares it to the intensity of light reflected from a reference material (Io)(such as a white tile). The ratio I/Io is called the "reflectance", and is usually expressed as a percentage (per cent R).

What are the Basic Parts of a UV Spectrophotometer?

The basic parts of a spectrophotometer are a light source, a holder for the sample, a diffraction grating in a mono-chromator or a prism to separate the different wavelengths of light, and a detector. A diffraction grating is an optical component with a periodic structure, which splits and diffracts light into several beams travelling in different directions. A monochromator is an optical device that transmits a mechanically selectable narrow band of wavelengths of light or other radiation chosen from a wider range of wavelengths available at the input.

The radiation source is often a Tungsten filament (300-2500 nm), a deuterium arc lamp, which is continuous over the ultraviolet region (190-400 nm), Xenon arc lamps, which is continuous from 160-2,000 nm; or more recently, light emitting diodes (LED) for the visible wavelengths. A deuterium arc lamp (or simply deuterium lamp) is a low-pressure gas-discharge light source often used in spectroscopy when a continuous spectrum in the ultraviolet region is needed. Gas-discharge lamps are a family of artificial light sources that generate light by sending an electrical discharge through an ionized gas. A xenon arc lamp is an artificial light source.

The detector is typically a photomultiplier tube, a photodiode, a photodiode array or a charge-coupled device (CCD). A photo multiplier tube is a type of gas-filled or vacuum tube that is extremely sensitive to light in the ultraviolet, visible, and near-infrared ranges of the electromagnetic spectrum. A photodiode is a type of photo detector capable of converting light into either current or voltage, depending upon the mode of operation. A charge-coupled device (CCD) is a device for the movement of electrical charge, usually from within the device to an area where the charge can be manipulated.

2.2.2. Different types of UV Spectrophotmeter

There are two principle types of uv-visible instruments. Filter photometers and Spectrophotometers

2.2.2.1. Filter Photometers

Filter photometers use filters to select the desired wavelength and are either single beam or double beam. A single beam instrument uses a single light path for both the reference and the sample. A double beam uses separate light paths for the reference and sample.

2.2.2.1.1. Single Beam Photometers

Single beam photometers use the same light path for both the solvent (100 per cent T) and the solvent-sample. Typically, the light passes through a collimating lens that is directed at the entrance slit where the light is then passed through a filter. The unabsorbed light is passed through the sample holder, passes a closable shutter (0 per cent T), to a photovoltaic cell which is attached to a readout device.

A stabilized power supply is very important. Using a stabilized power supply avoids errors resulting from changes in the beam intensity during the time required to measure 100 per cent T and the per cent T of the analyte.

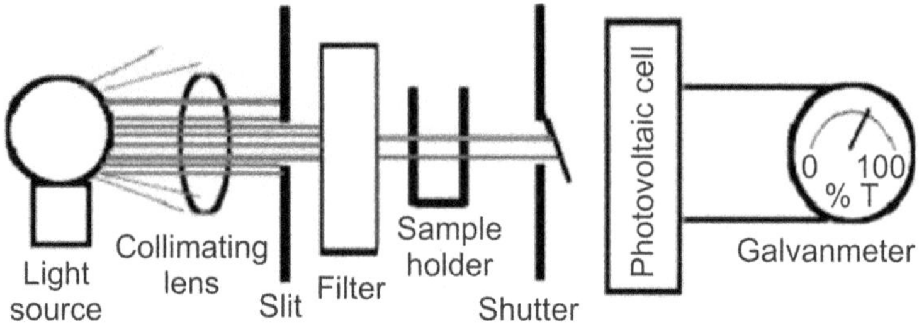

Advantages of Single Beam Photometers

- ☆ Low cost
- ☆ Ease of operation

Disadvantages of Single Beam Photometers

- ☆ Variation in light intensity — errors in per cent T
- ☆ Light beam not monochromatic — deviation from Beer's law
- ☆ Per cent T and A are not "true" values
- ☆ Not designed for spectral data

2.2.2.1.2. Double Beam Photometers

Double beam photometers split the light path into two segments. One segment is used as a reference and the other as the working path. This allows the instrument to make automatic corrections for variations in light intensity.

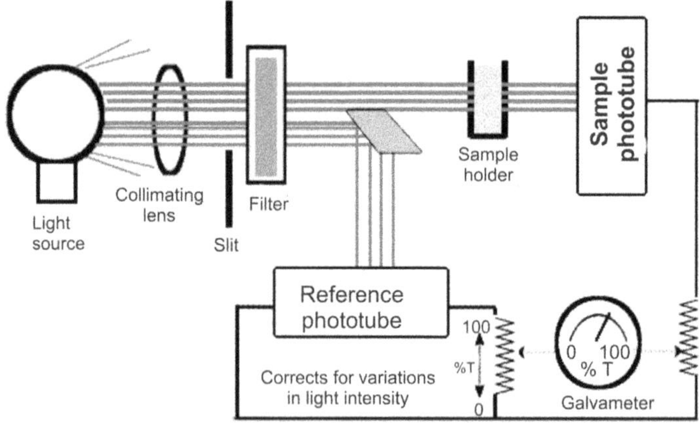

When the voltages between the sample resistor (CD) and reference resistor (AB) are equal, no current flows through the galvanometer (null detector) since there is no voltage potential difference between the two resistors.

Advantages of Double Beam Photometers

- ☆ Correction for changes in light intensity
- ☆ Ease of operation

Disadvantages of Single Beam Photometers

- ☆ Light beam is not monochromatic — deviation from Beer's law
- ☆ Per cent T and A are not "true" values
- ☆ Not designed for spectral data

2.2.2.2. Spectrophotometers

Spectrophotometers use monochromatic dispersion elements to vary the wavelengths. continuously variable wavelength selection and light source is UV or VIS.

2.2.2.2.1. Optical Path of Single Beam Spectrophotometers

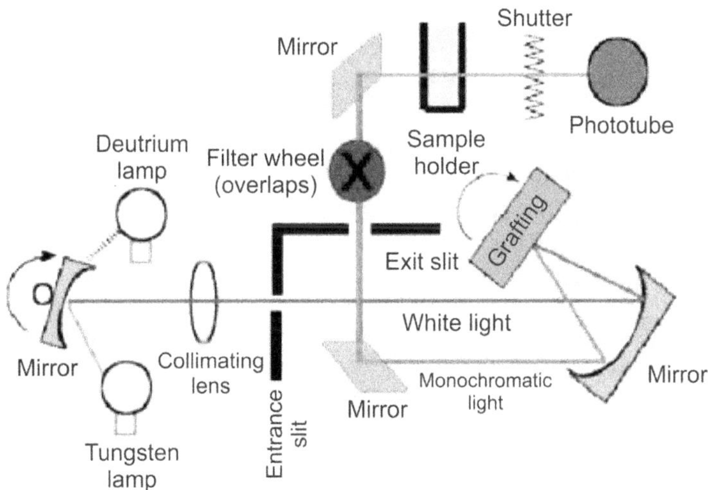

Electronics of Single Beam Spectrophotometers

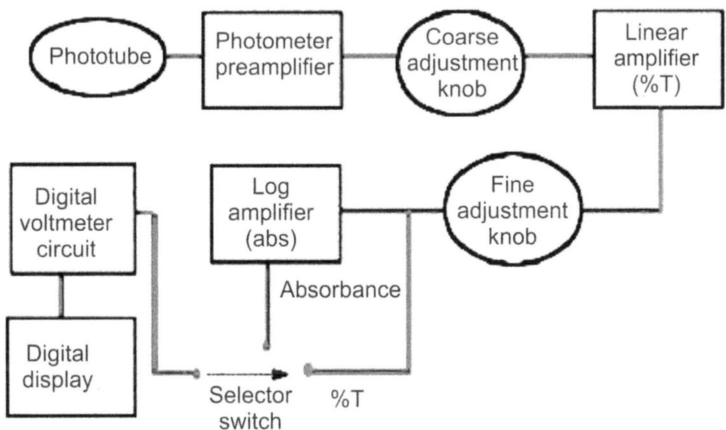

Advantages of Single Beam Spectrophotometer

- ☆ Wavelengths easily selected
- ☆ Low cost

Disadvantages of Single Beam Spectrophotometer

- ☆ Cannot easily be used for absorption spectra (point by point).
- ☆ Changes in light intensity cause variations in readout.
- ☆ Changes in solvent causes variations.
- ☆ Changes in wavelengths cause variations and per cent T and has to be reset at 0 per cent T and 100 per cent T.

2.2.2.2.2. Double Beam Spectrophotometers

Double beam spectrophotometers use a beam chopper to separate the reference beam from the sample beam.

Beam Chopper

The chopper is divided into three equal segments. One segment is clear which will pass light through the wheel, one segment is mirrored which will reflect light along a different path, and the third segment is opaque so that no light is transmitted nor reflected.

Advantages of Double Beam Spectrophotometers

- ☆ Speed of operation
- ☆ Automatic compensation for variation in lamp output
- ☆ Solvent absorption at various solution
- ☆ Changes in detector sensitivity
- ☆ Spectra scan

Disadvantages of Double Beam Spectrophotometers

- ☆ Cost is higher

2.2.3. Components of UV/Vis Spectrophotometer

Samples for **UV/Vis spectrophotometer** are most often liquids, although the absorbance of gases and even of solids can also be measured. Samples are typically placed in a transparent cell, known as a cuvette. Cuvettes are typically rectangular in shape; commonly with an internal width of 1 cm. Cuvettes are made of high quality fused silica or quartz glass because these are transparent throughout the UV, visible and near infrared regions.

Instruments for measuring the absorption of U.V. or visible radiation are made up of the following components;

- ☆ Sources (UV and visible)
- ☆ Wavelength selector (monochromator)
- ☆ Sample containers
- ☆ Detector
- ☆ Signal processor and readout
- ☆ Each of these components will be considered in turn.

2.2.3.1. Light Source

The light source is a tungsten filament bulb for the visible part of the spectrum and a deuterium bulb for the UV part of the spectrum. A mechanically moving mirror change the light source from visible to UV region at about 350 nm and directs appropriate beam of light along the optical axis. The emitted light consists of many different wavelengths. It is important that the power of the radiation source does not change abruptly over it's wavelength range. The electrical excitation of deuterium or hydrogen at low pressure produces a continuous UV spectrum. The mechanism for this involves formation of an excited molecular species, which breaks up to give two atomic species and an ultraviolet photon. Both deuterium and hydrogen lamps emit radiation in the range 160 - 375 nm. Quartz windows must be used in these lamps, and quartz cuvettes must be used, because glass absorbs radiation of wavelengths less than 350 nm.

The tungsten filament lamp is commonly employed as a source of visible light. This type of lamp is used in the wavelength range of 350 - 2500 nm. The energy emitted by a tungsten filament lamp is proportional to the fourth power of the operating voltage. This means that for the energy output to be stable, the voltage to the lamp must be very stable indeed. Electronic voltage regulators or constant-voltage transformers are used to ensure these stability. Tungsten/halogen lamps contain a small amount of iodine in a quartz "envelope" which also contains the tungsten filament. The iodine reacts with gaseous tungsten, formed by sublimation, producing the volatile compound WI2. When molecules of WI2 hit the filament they decompose, redepositing tungsten back on the filament. The lifetime of a tungsten/ halogen lamp is approximately double that of an ordinary tungsten filament lamp. Tungsten/halogen lamps are very efficient, and their output extends well into the ultra-violet. They are used in many modern spectrophotometers.

2.2.3.2. Monochromators

A monochromator is placed between the light source and the sample. It is an optical system which produces radiation of a single wavelength. It consists of either a prism or a rotating metal grid of high precision called gratings. If wavelengths are selected by prisms or gratings, it is known as spectrophotometry.

Diagram of Monochromator

Prisms are made of quartz or silica because glass absorbs wavelength less

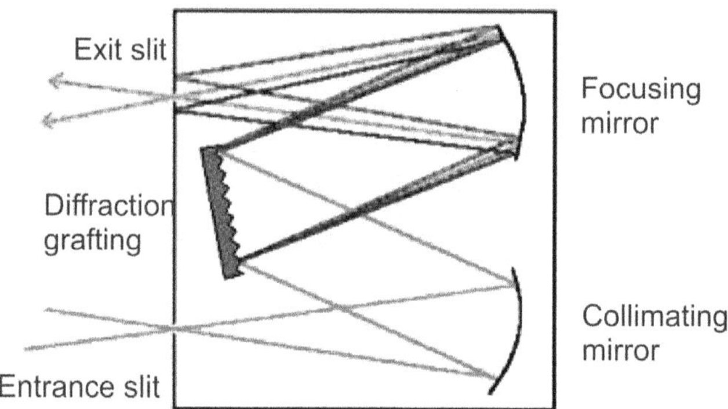

than 400 nm. Rotation of gratings or prisms perpendicular to the optical axis helps to select the different wavelengths. A prism splits the incoming light into its components by refraction. Refraction occurs as radiation of different wavelength travels along different paths. A grating splits the wavelengths by diffraction. Diffraction is a reflection phenomenon that occurs at a grid surface or engraved fine lines. Diffraction of light gives rise to a series of overlapping spectra.

The bandwidth is a function of the optical slit width. The narrower the slit width the more reproducible is the absorbance values. On contrary, the sensitivity becomes less as the slit narrows because less radiation passes through to the detector. The resolution achieved by gratings is much higher than prisms. Gratings give high **resolving power** and it is directly proportional to the closeness of lines. Gratings can resolve lower wavelengths more than higher ones.

Polychromatic radiation (radiation of more than one wavelength) enters the monochromator through the entrance slit. The beam is collimated, and then strikes the dispersing element at an angle. The beam is split into its component wavelengths by the grating or prism. By moving the dispersing element or the exit slit, radiation of only a particular wavelength leaves the monochromator through the exit slit.

Czerney-Turner Grating Monochromator

All monochromators contain the following component parts;

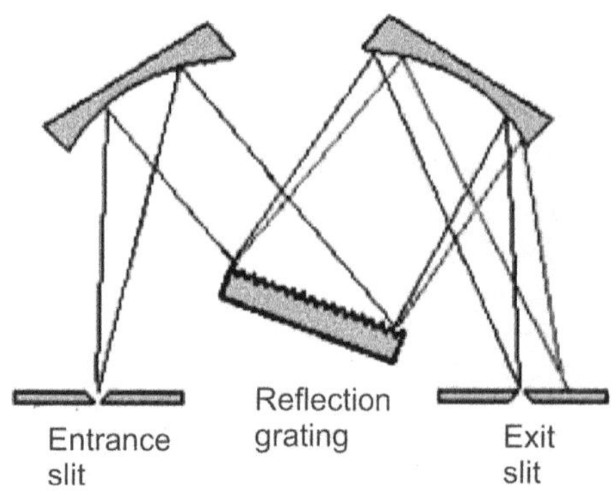

- An entrance slit
- A collimating lens
- A dispersing device (usually a prism or a grating)
- A focusing lens
- An exit slit

2.2.3.3. Sample cell

Sample dissolved in suitable solutions is taken in an optically transparent sample cell called cuvette. Borosilicate glass and normal plastics cuvettes are used in the visible range of spectrum (upto 350 nm) as they absorb UV light. Quartz or silica cuvettes are used for UV range of spectrum and are transparent to wavelength above 180 nm. A control or reference cuvette should be optically identical to the test cuvette. The optical path length of the cuvette is mostly 1 cm. Cuvettes with path length of 100, 20, 5, 2 and 1 mm are also available for particular applications. Disposable plastic cuvettes are now available that allow measurements over the entire range of the UV/Vis spectrum. The containers for the sample and reference solution must be transparent to the radiation which will pass through them. Quartz or fused silica cuvettes are required for spectroscopy in the UV region. These cells are also transparent in the visible region. Silicate glasses can be used for the manufacture of cuvettes for use between 350 and 2000 nm.

2.2.3.4. Detectors

Detectors used in spectrophotometers are either photocells or photomultiplier cells. Photocells convert quanta of radiation into electrical energy, which is then amplified, detected and recorded. Photons impinging on **metal surface** in a vacuum cause emission of electrons in proportion to the intensity of radiation or light. A positive electrode attracts these emitted electrons and causes a current flow and this gives a potential difference across a resistor. This potential is electronically amplified and the light absorbed is measured. Photocells are sensitive to light with a wavelength of about 400 nm and insensitive to wavelengths >550 nm. Theoretically accuracy of photocells is 1±0.003.

Photomultiplier tube (PMT) is more sensitive than photocells. It consists of a photocathode and dynodes in an evacuated chamber. Photons that strike the

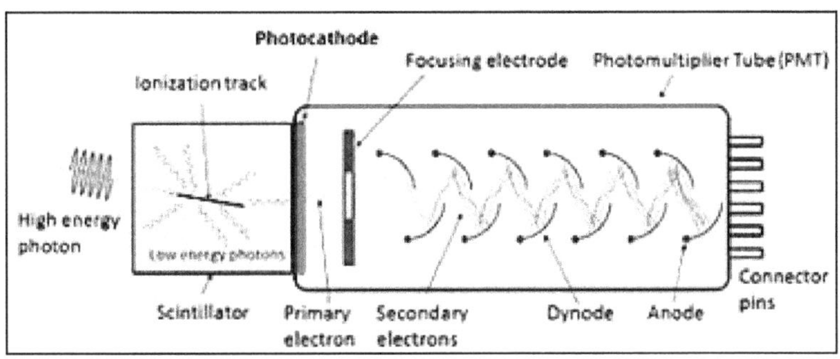

photo emissive cathode emit electrons due to the photoelectric effect. Electrons are accelerated towards a series of additional electrodes called dynodes. Additional electrons are generated at each dynode. This cascading effect creates 105 to 107 electrons for each photon. This amplified signal is finally collected at the anode and measured. The illustration of PMT can be viewed.

It is better to have one detector instead of two detectors to measure the intensity of light from the control and test samples, simultaneously. This helps to eliminate the potential variations between the two detectors.

Photomultipliers are very sensitive to UV and visible radiation. They have fast response times. Intense light damages photomultipliers; they are limited to measuring low power radiation.

Cross Section of a Photomultiplier Tube

The linear photodiode array is an example of a multichannel photon detector. These detectors are capable of measuring all elements of a beam of dispersed radiation simultaneously. A linear photodiode array comprises many small silicon photodiodes formed on a single silicon chip. There can be between 64 to 4096 sensor elements on a chip, the most common being 1024 photodiodes. For each diode, there is also a storage capacitor and a switch. The individual diode-capacitor circuits can be sequentially scanned. In use, the photodiode array is positioned at the focal plane of the monochromator (after the dispersing element) such that the spectrum falls on the diode array. They are useful for recording UV-Vis. absorption spectra of samples that are rapidly passing through a sample flow cell, such as in an HPLC detector.

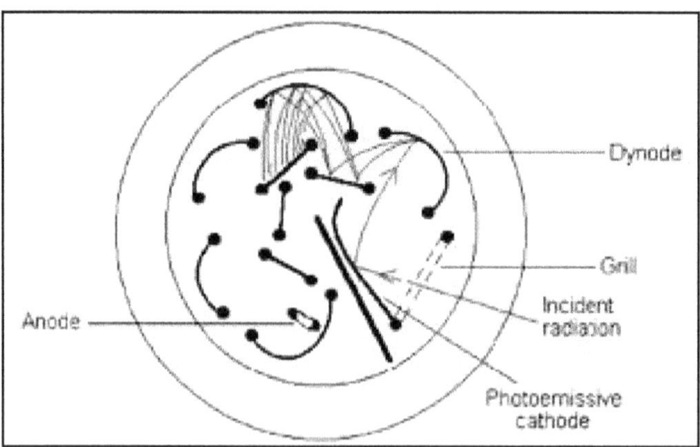

Charge-Coupled Devices (CCDs) are similar to diode array detectors, but instead of diodes, they consist of an array of photocapacitors.

2.2.3.5. Applications

The usual procedure for colorimetric assays is to prepare a set of standards and produce a plot of concentration versus absorbance called calibration curve.

This should be linear. Absorbances of unknown sample are then measured and their concentration interpolated from the linear region of the plot. The maximum absorbance should be approximately 0.5 to obtain good spectra.

A UV/Vis spectrophotometer may be used as a detector for HPLC which is the High-performance liquid chromatography (or high-pressure liquid chromatography, HPLC) is a chromatographic technique that can separate a mixture of compounds and is used in biochemistry and analytical chemistry to identify, quantify and purify the individual components of the mixture.

UV/Vis spectroscopy is routinely used in analytical chemistry for the quantitative determination of different analytes, such as transition metal ions, highly conjugated organic compounds, and biological macromolecules. Determination is usually carried out in solutions. Transition metals are elements which belong to groups 3 to 12 on the periodic table. Conjugated system is a system with delocalized electrons (electrons in a molecule or solid metal that are not associated with a single atom) in compounds with alternating single and multiple bonds

Qualitative analyses are performed in the UV/Vis region to identify certain classes of compounds in pure state as well as in biological mixtures.

Quantification of biological samples either directly or via colorimetric assays is most common. Proteins are quantified directly using their intrinsic chromophores, tyrosine and tryptophan.

Difference spectroscopy detects small absorbance changes in systems with high background absorbance. A difference spectrum is obtained by subtracting one absorption spectrum from another. Common applications are determination of number of aromatic amino acids exposed to solvent, detection of conformational changes occurring in proteins, detection of aromatic amino acids in active sites of enzymes and monitoring of reactions involving catalytic chromophores.

Derivative spectroscopy differentiates the absolute absorption spectrum of a sample and differentially plot against the wavelength. The successful applications include the binding of a monoclonal antibody to its antigen with second order derivatives and quantification of tryptophan and tyrosine residues in proteins with fourth order derivatives.

Biochemical assays are performed based on the continuous monitoring of the absorbance of a system at a given wavelength throughout the course of the experiment. It is used for enzyme assays and kinetic analysis.

Chapter 3

Chromatography

3.1. Basic Principles of Chromatographic Techniques

Chromatography techniques are used for the separation of one or more biological compounds from a mixture. The compounds are separated in large as well as small amounts (picograms) by this technique. The term 'chromatography' was coined by Mikhail Tswett in 1906. "Chroma" means color and "graphein" means written. The name was coined so as initially, this technique is used for the separation of coloured plant pigments and later applied to colourless compounds. All chromatographic system consists of two phases

- ☆ Stationery phase: which may be solid, gel, liquid supported on a solid
- ☆ Mobile phase: which may be liquid or gas and flows through the stationery phase

Separations of the compounds are mainly based on the differential interaction of molecules between a mobile phase and a stationary phase. Selection of a chromatographic technique is made such that the compounds to be isolated have different distribution coefficients.

Chromatography is a non-destructive procedure for resolving a complex mixture into its individual fractions or compounds. It is a separation procedure, and the separated entities are identified by other analytical techniques like UV-visible, Infrared, NMR (nuclear magnetic resonance), Mass spectroscopy, and so forth. For a quantitative analysis, measurements of the area under the curve in the chromatogram are taken. Its name is derived from two words: "chromo" meaning colour, and "graphy" meaning writing. In other words, colour bands are formed in the procedure, which are then measured or analysed. These bands are due to separation of individual compounds at different lengths on the column, as seen in column chromatography and on paper in paper chromatography. However, in modern methods like HPLC or gas chromatograph, colour bands can't be seen. The basic principle of chromatography has advanced a lot to cater to the growing needs of the industry and for research purposes.

The samples are subjected to flow by mobile liquid onto or through the stable stationary phase. The sample components are separated into fractions based on their relative affinity towards the two phases during their travel. The fraction with a greater affinity to stationary layer travels slower and at a shorter distance, while that with a lesser affinity travels faster and longer.

3.1.1. Chromatographic Separation

Distribution Coefficients

The basis of all forms of chromatography is the distribution or partition coefficient (Kd), which describes the way in which a compound (the analyte) distributes between two immiscible phases. For two such phases A and B, the value for this coefficient is a constant at a given temperature and is given by the expression:

Kd = concentration in phase A/concentration in phase B

The term effective distribution coefficient is defined as the total amount, as distinct from the concentration, of analyte present in one phase divided by the total amount present in the other phase. It is in fact the distribution coefficient multiplied by the ratio of the volumes of the two phases present. If the distribution coefficient of an analyte between two phases A and B is 1, and if this analyte is distributed between 10 cm3 of A and 1 cm3 of B, the concentration in the two phases will be the same, but the total amount of the analyte in phase A will be 10 times the amount in phase B. All chromatographic systems consist of the stationary phase, which may be a solid, gel, liquid or a solid/liquid mixture that is immobilised, and the mobile phase, which may be liquid or gaseous, and which is passed over or through the stationary phase after the mixture of analytes to be separated has been applied to the stationary phase. During the chromatographic separation the analytes continuously pass back and forth between the two phases so that differences in their distribution coefficients result in their separation.

Adsorbtion: Interaction of solute molecules (or atoms or ions) with the surface of the stationary phase (note that it is different from absorption where the molecules fill the pores of a solid).

- ☆ **Eluent:** The mobile phase (usually for solvents)
- ☆ **Elution:** Motion of the mobile phase through the stationary phase
- ☆ **Elution time:** The time taken for a solute to pass through the system. A solute with a short elution time travels through the stationary phase rapidly, *i.e.* it elutes fast.

- ☆ **Mobile phase:** The part of the chromatography system that is mobile. Commonly a solvent mixture (as in column chromatography or thin layer chromatography or a gas (as in gas chromatography).
- ☆ **Normal phase:** "Unmodified" stationary phase where POLAR solutes interact strongly and run slowly Reverse phase: "Modified" stationary phase where POLAR solutes run fast *i.e.* reverse order
- ☆ **Resolution:** Degree of separation of different solutes. In principle, resolution can be improved by using a longer stationary phase, finer stationary phase (*e.g.* column packing or TLC plate coating) or slower elution.
- ☆ **Stationary phase:** The part of the chromatography system that is fixed in place. Most commonly a solid *e.g.* the packing in column chromatography or gas chromatography or the coating on a chromatographic plate.

3.1.2. Types of Chromatographic System

Types

(1) Based on the **technique** employed in separation of components it is classified

Adsorption-based

Here, the stationary layer is a solid while the mobile phase is liquid. The compounds travel on the solid surface under the influence of mobile liquid. The separation depends on the extent of physical adsorption to the solid surface.

Partition-Based

In this mode, both the stationary and mobile phases are liquids. So the compounds have an affinity based on their partition coefficients into the individual liquid layer. The one with the greater partition to mobile liquid has a higher affinity to it, so it travels faster, and vice versa.

(2) Based on the type of stationary **material** used for separation, there are two types:

Normal Phase

Here, the stationary material is polar in nature and hence the compounds with a higher polarity elute out last while non polars come out first.

Reverse Phase

Here, the stationary material is non-polar in nature and hence the compounds with a lower polarity elute out last, and vice versa.

In most HPLC analyses, the type used is the reverse one, as many of the biological, phytochemical compounds, and drugs estimated by HPLC are polar in nature.

Techniques

There are many developments which have occurred over years based on the requirements and also technology employed in evaluation of mixtures. The techniques can be broadly divided into **planar** and **columnar** techniques.

Planar

In this type, the stationary phase is a plane surface (two dimensions surface where only length and breadth are taken as area), on which chromatograms are formed.

This method is adopted in techniques like

1. Using papcr
2. Thin layer chromatography
3. High performance thin layer chromatography (HPTLC).

Columnar

In this type, there is use of a column wherein on its walls lies a stationary phase, while the mobile phase is flushed through the column.

The techniques which employ this method are:

1. Column chromatography
2. Gas chromatography
3. HPLC
4. Size exclusion
5. Ion exchange

Pros and Cons of Both Techniques

Both techniques have their pros and their cons.

1. Planar techniques have the advantages of faster separation, visualization of formed chromatograms, or spots, and they are less expensive. However they are not useful for preparative purposes.
2. Columnar techniques have the advantages of having better or more effective separations of even complex mixtures — possibility to yield to a large amount of compounds by separation of mixtures, i.e in preparative mode. However they have the disadvantage of being expensive, time consuming, and cumbersome (heavy).

3.1.3. Equilibrations

A theoretical plate in many separation processes is a hypothetical zone or stage in which two phases, such as the liquid and vapor phases of a substance, establish equilibrium with each other. Such equilibrium stages may also be referred to as an equilibrium stage, ideal stage, or a theoretical tray. The performance of many separation processes depends on having a series of equilibrium stages and is enhanced by providing more such stages. In other words, having more theoretical plates increases the efficiency of the separation process be it either a distillation, absorption, chromatographic, adsorption or similar process.

Methods for Describing Column Efficiency

There are two related terms used as quantitative measures of chromatographic efficiency:

Number of 'theoretical plates' - N

Plate height - H

These parameters are used to measure a column's resolving power (ability to differentiate between closely migrating peaks - sometimes called efficiency). The two parameters are related by the following equation:

$$N = 5.545\left(\frac{t_R}{w^h}\right)^2 \qquad (4)$$

N: Number of theoretical plates

t_R: Retention time

w_h: Peak width at half height (in units of time)

Number of Theoretical Plates (N), Height Equivalent to a Theoretical Plate (H), Utilization of Theoretical Efficiency (UTE per cent), Resolution (RS), Phase Ratio (β), Number of Theoretical Plates (N) Also known as column efficiency, the number of theoretical plates is a mathematical concept and can be calculated using Equation. A capillary column/any chromatography column does not contain anything resembling physical distillation plates or other similar features. Theoretical plate numbers are indirect measure of peak width for a peak at a specific retention time. Columns with high plate numbers are considered to be more efficient, that is, have higher column efficiency, than columns with a lower plate count. A column with a high number of theoretical plates will have a narrower peak at a given retention time than a column with a lower N number. High column efficiency is beneficial since less peak separation (meaning lower alpha, α) is required to completely resolve narrow peaks. On stationary phases where the alphas (α) are small, more efficient columns are needed. Column efficiency is a function of the column dimensions (diameter, length and film thickness), the type of carrier gas and its flow rate or average linear velocity, and the compound and its retention. For column comparison purposes, the number of theoretical plates per meter (N/m) is often used.

Theoretical plate numbers are only valid for a specific set of conditions. Specifically, isothermal temperature conditions are required because temperature programs result in highly inflated, inaccurate plate numbers. Also, the retention factor (k) of the test solute used to calculate plate numbers should be greater than 5. Less retained peaks result in inflated plate numbers. When comparing theoretical plate numbers between columns, the same temperature conditions and peak retention (k) are required for the comparison to be valid.

Height Equivalent to a Theoretical Plate (H)

Another measure of column efficiency is the height equivalent to a theoretical plate denoted as H. It is calculated using Equation and usually reported in millimeters. The shorter each theoretical plate, the more plates are "contained" in any length of column. This, of course, translates to more plates per meter and higher column efficiency.

$$H = \frac{L}{N}$$

L: Length of column (mm)

N: Number of theoretical plates

3.1.4. Applications

Chromatographic techniques are used in different applications such as:

1. Separation of pigments, proteins, amino acids, DNA and RNA.
2. Separation of colourless substances from a complex mixture.
3. Fractionation of gases, liquids or dissolved solids.
4. Biomolecules are purified using chromatography techniques that separate them according todifferences in their specific properties
   - ✰ Size - Gel filtration (GF), also called size exclusion
   - ✰ Charge - Ion exchange chromatography (IEX), chromatofocusing (CF)
   - ✰ Hydrophobicity - Hydrophobic interaction chromatography (HIC) and Reversed phase chromatography (RPC)
   - ✰ Biorecognition (ligand specificity) - Affinity chromatography (AC)

3.2. Gel Filtration

3.2.1. Gel Filtration Chromatography

Gel filtration chromatography is also known as size exclusion chromatography or permeation chromatography or molecular sieve chromatography. In this technique, the separation of molecules is achieved based on their molecular size. The molecular sieve properties of a variety of porous gel materials are utilized for the separation of molecules. The dextran beads are first porous gel materials introduced for the separation of biological molecules such as polysaccharides, proteins, peptides and glycoproteins.

In gel filtration chromatography, the stationary phase consists of porous beads with a well-defined range of pore sizes. The stationary phase for gel filtration is said to have a fractionation range, meaning that molecules within that molecular weight range can be separated. Proteins that are small enough can fit inside all the pores in the beads and are said to be included. These small proteins have access to the mobile phase inside the beads as well as the mobile phase between beads and elute last in a gel filtration separation.

Proteins that are too large to fit inside any of the pores are said to be excluded. They have access only to the mobile phase between the beads and, therefore, elute first. Proteins of intermediate size are partially included - meaning they can fit inside some but not all of the pores in the beads. These proteins will then elute between the large ("excluded") and small ("totally included") proteins. Consider the separation of a mixture of glutamate dehydrogenase (molecular weight 290,000), lactate dehydrogenase (molecular weight 140,000), serum albumin (MW 67,000), ovalbumin (MW 43,000), and cytochrome c (MW 12,400) on a gel filtration column packed with Bio-Gel P-150 (fractionation range 15,000-150,000). When the protein mixture is applied to the column, glutamate dehydrogenase would elute first because it is above the upper fractionation limit. Therefore it is totally excluded from the inside of the porous stationary phase and would elute with the void volume (V_0). Cytochrome c is below the lower fractionation limit and would be completely included, eluting last. The other proteins would be partially included and elute in order of decreasing molecular weight.

Principle

Gel filtration is performed in a column having porous matrix support. The matrix support consists of porous gel beads or glass granules. The column is kept in equilibrium with the solvent mobile phase. The matrix support has two measurable liquid volumes. They are external (Vo) and internal (Vi) volumes. External or void volume means the volume of liquid occupying the spaces between the packing matrix beads. Internal volume means the volume occupied by the liquid within the pores of the matrix beads. The larger molecules do not enter the pores of the matrix beads and pass out quickly through the column. The smaller molecules enter the pores of the matrix beads and their flow is retarded as they equilibrate with the internal and external volumes. Their extent of availability in the pores depends on the porosity of the matrix beads and size of the molecules.

Total volume is the volume occupied by the matrix beads in the column and not the size of the beads. Partition coefficient defines the proportion of the bead pores that can be occupied by the molecule.

The molecular size of an unknown molecule can be determined by comparing the elution volume of the molecules with known molecular weights.

These separations can be described by this equation:

$$\mathbf{V_r = V_o = KV_i}$$

Where V_r is the retention volume of the protein, V_0 is the volume of mobile phase between the beads of the stationary phase inside the column (sometimes called the void volume), V_i is the volume of mobile phase inside the porous beads (also called the included volume) and K is the partition coefficient (the extent to which the protein can penetrate the pores in the stationary phase, with values ranging between 0 and 1).

In the mixture of proteins listed above, the partition coefficient (K) for glutamate dehydrogenase would be 0 (totally excluded), K = 1 for cytochrome c (totally included) and K would be between 0 and 1 for the other proteins, which are within the fractionation range for the column.

In practice, gel filtration can be used to separate proteins by molecular weight at any point in a purification of a protein. It can also be used for buffer exchange - a protein dissolved in a sodium acetate buffer, pH 4.8, can be applied to a gel filtration column that has been equilibrated with tris buffer, pH 8.0. Using the tris buffer, pH 8.0, as the mobile phase, the protein moves into the tris mobile phase as it travels down the column, while the much smaller sodium acetate buffer molecules are totally included in the porous beads and travels much more slowly than the protein.

Selection of Operating Conditions

Various factors should be considered when designing a gel-filtration system. These include: (i) matrix choice; (ii) sample size and concentration; (iii) column parameters; (iv) choice of eluent; (v) effect of flow rate, and (vi) column cleaning and storage.

3.2.2. Matrix Choice

Commonly used gel-filtration matrices consist of porous beads composed of cross linked polyacrylamide, agarose, dextranor combinations of these, and are supplied either in suspended form or as dried powders. The matrix should be compatible with the properties of the molecules being separated and its stability to organic solvents, pH and temperature is also an important consideration. Under separation conditions, matrices should be inert with respect to the molecules being separated in order to avoid partial adsorption of the molecules to the matrix, not only retarding their migration through the column, but also resulting in "tailed" peaks.

When choosing a suitable matrix, one with a molecular mass fractionation range which will allow the molecule of interest to elute after V0 and before Vt, should be selected. The most suitable fractionation range, however, will be dictated not only by the molecular mass of the target molecule, but also by the composition of the sample being applied to the column. Therefore, the best separation of molecules within a sample having similar molecular masses is achieved using a matrix with a narrow fractionation range.

Maximum resolution in gel-filtration chromatography depends on application of thesample in a small volume, typically 1-5 per cent of the total bed volume. For this reason, gel-filtration chromatography has an inherent low sample-handling capacity and accordingly, should be performed quite late in a purification procedure when thenumbers of different molecules in a sample are relatively low. The concentration of sample which can be applied to the column will be limited by the viscosity of the sample (which increases with sample concentration) relative to that of the eluent. A high viscosity will result in irregular sample migration through the column with subsequent loss of resolution and, in some instances, will reduce the column flowrate. When separating proteins by gel-filtration, the sample should not have a protein concentration in excess of 20 mg/ml.

3.2.3. Column Parameters

Maximum resolution in gel-filtration chromatography is obtained with long columns.

The ratio of column diameter to length can range from 1:20 up to 1:100.

3.2.4. Choice of Eluent

As gel-filtration chromatography separates molecules only on the basis of their relative sizes, the technique is effectively independent of the type of eluent used. Elution conditions (pH, essential ions, cofactors, protease inhibitors *etc.*) which will complement the requirements of the molecule of interest should, therefore, be selected. However, the ionic strength of the eluent should be high enough to minimize protein-matrix and protein-protein associations by electrostatic or van der Waals interactions. The addition of 0.1M NaCl or KCl to the eluent to avoid these interactions is quite common.

3.2.5. Effect of Flow Rate

Low flow rates offer maximum resolution during gel-filtration chromatography; since flow rate and resolution are inversely related. The optimum flow rate for resolution of proteins is approximately 2 $ml/cm^2/h$, although much higher flow rates can be used, particularly with rigid matrices such as the Sephacryl HR range from GE Healthcare (30 $ml/cm^2/h$). Unfortunately, low flow rates mean longer separation times. Therefore, a compromise between desired resolution and speed must be decided upon.

3.2.6. Column Cleaning and Storage

Most gel-filtration matrices can be cleaned with 0.2 M sodium hydroxide or non-ionic detergents. When left unused for long periods of time, matrices should be stored at 4°C in the dark in the presence of an antimicrobial agent (0.02 - 0.05 per cent w/v sodiumazide or 20 per cent v/v ethanol).

3.2.7. Column Matrices

The media used for gel exclusion chromatography include dextran (Sephadex), polyacrylamide (Bio-Gel P) and dextran-polyacrylamide (Sephacryl) and agarose (Sepharose and BioGel A). Each is available with a variety of different ranges of pore size in the beads, permitting separation of macromolecules of different size.

The table shows the useful range for the most commonly used gel filtration media the lower and upper molecular sizes (in kDa) over which they can be used to separate macromolecules. The upper limit is known as the exclusion limit of the gel - the size above which proteins will elute in the void volume of the column.

	Lower Limit	*Upper Limit*
Dextran gels		
Sephadex G-50	1.5	30
Sephadex G-75	3	80
Sephadex G-100	4	150
Sephadex G-150	5	300
Sephadex G-200	5	600
Polyacrylamide gels		
Bio-Gel P-10	1.5	20
Bio-Gel P-30	2.5	40
Bio-Gel P-60	3	60
Bio-Gel P-100	5	100
Bio-Gel P-150	15	150
BioGel P-200	30	200
Bio-Gel P-300	60	400
Dextran-polyacrylamide gels		
Sephacryl S-200	5	250
Sephacryl S-300	10	1500
Sephacryl S-400	20	8000
Agarose gels		
Sepharose 6B	10	4000
Sepharose 4B	60	20,000
Sepharose 2B	70	40,000
Bio-Gel A-0.5	10	500
Bio-Gel A-1.5	10	1500
Bio-Gel A-5	10	5000
Bio-Gel A-15	40	15,000
Bio-Gel A-50	100	50,000

Support matrices range from soft low pressure beaded gels to rigid high performance silica porous glass granules. Common beaded gels include cross-linked dextrans, agarose, polyacrylamide and polystyrene gels. Porous glass granules are bio-glass and porous silica. Majority of gels are available in two particle size grades *viz.* coarse (300-100 mm) and fine (20-80 mm).

3.2.7.1. Dextran Gels

The dextran gel materials are most common support matrices available in dry powder form and commercially known as 'Sephadex'. Dextran gel is made by

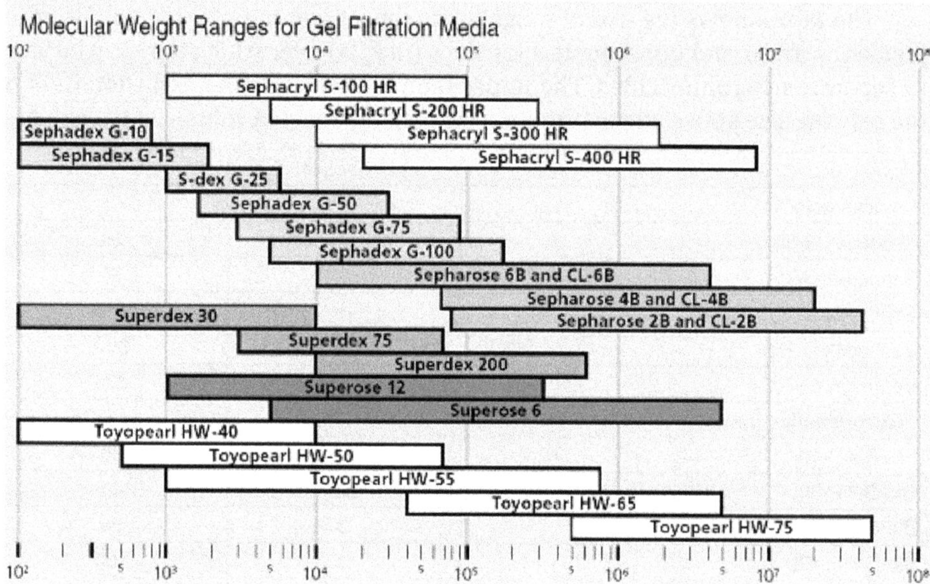

cross-linking a polysaccharide, dextran with epichlorohydrin. They have hydrophilic character and swells in aqueous media. They are stable in water, salt solution, organic solvents, alkali and weak acidic solution. They are easily susceptible to bacterial contamination and hence require an addition of preservative during storage. Sephadex are available as a series of beads with different separation ranges.

Gel Type	*Fractionation Range (Da)*
G-10	100–800
G-15	500–1,500
G-25	1000–5,000
G-50	1,500–30,000
G-75	3,000–80,000
G-100	4,000–1,50,000

3.2.7.2. Agarose Gels

Agarose gels are popular gel materials available in beaded form. They consist of a polysaccharide, agar with D-galactose and 3,6 anhydro L-galactose units and commercially available as Bio-Gel A, Ultragel and Sepharose. These gel materials are hydrophilic but have no charged groups. They are compatible with cell aqueous buffers and stable at pH 4-10. They are more rigid than dextran gels. They allow separation of high molecular weight molecules and subcellular fractions. They are mostly used to study viruses, nucleic acids, polysacharrides and membrane proteins. They are supplied as dense suspension of beads incorporated with preservative, 0.2 per cent sodium azide. The fractionation ranges of agarose based gels are given in Table.

Gel Type

Fractionation	Range(Da)
	Bio-Gel
A-0.5m	10,000-50,000
A-1.5m	10,000-1,500,000
A-5m	10,000-5,000,000
A-15m	40,000-15,000,000
A-50m	100,000-50,000,000
A-150M	1,000,000-150,000,000
	Sepharose
2B	70,000-40,000,000
4B	60,000-20,000,000
6B	10,000-4,000,000
	Ultra gel
A2	120,000-50,000,000
A4	55,000-9,000,000
A6	25,000-2,400,000

3.2.7.3. Polyacrylamide Gels

Polyacrylamide gels are formed by polymerization of acrylamide and methylene bis acrylamide. They are stable in aqueous buffers at pH 1-10. They withstand higher pressures and have lower susceptibility to bacterial contamination. They are supplied as a dry powder and swollen prior to use. They are commercially available as Bio-Pel P gels. The fractionation ranges of polyacrylamide gels are given in Table.

Gel Type	Fractionation Range (Da)
P 2	100–1,800
P 4	800–4,000
P 6	1,000 – 6,000
P 10	1,500 – 20,000
P 30	2,500 – 40,000
P 60	3,000 – 60,000
P 100	5,000 – 1,00,000
P 200	30,000 – 2,00,000
P 300	60,000 – 4,00,000

3.2.7.4. Composite Gels

Sephracyl is a composite gel which consists of allyl dextran cross-linked with N,N'methylene bis-acrylamide. They are stable and possess a uniform pore size. They are more rigid to withstand high flow rates and high operational pressure. They are used for the separation of high molecular weight components. Ultragel AcA is another composite gel made by cross linking polyacrylamide with agarose to form a semi-rigid beaded packing material. They also posses uniform pore size. Composite gels are supplied as a suspension in buffer containing preservative. The fractionation range of the composite gels are given in Table.

Gel Type	*Fractionation Range (Da)*
Sephacryl	
S-200	5,000-250,000
S-300	10,000-1,500,000
S-400	20,000-8,000,000
S-500	40,000-20,000,000
S-1000	500,000-100,000,000
Ultra gel	
AcA 22	100,000-1,200,000
AcA 34	20,000-350,000
AcA 44	10,000-130,000
AcA 54	5,000-70,000
AcA 202	1,000 – 15,000

3.2.8. Chromatography Equipment

Gel permeation chromatography (GPC) is conducted almost exclusively in chromatography columns. The experimental design is not much different from other techniques of liquid chromatography. Samples are dissolved in an appropriate solvent, in the case of GPC these tend to be organic solvents and after filtering the solution it is injected onto a column. The separation of multi-component mixture takes place in the column. The constant supply of fresh eluent to the column is accomplished by the use of a pump. Since most analytes are not visible to the naked eye a detector is needed. Often multiple detectors are used to gain additional information about the polymer sample. The availability of a detector makes the fractionation convenient and accurate.

3.2.8.1. Columns

Columns are made of glass or plastic. Simple conventional column consists of single glass tubes equipped with rubber stopper and syringe needle at one end. Sophisticated columns are also made of glass tube with molded end fittings called as flow adaptors. Simple columns are run in a descending mode, while sophisticated columns are run in a ascending mode. Most efficient separations could be achieved only when the sample is pumped up through the separation bed in ascending mode. Commercial columns are usually 7-7.5 mm in diameter and 30-60 cm in length.

3.2.8.2. Gel

Gels are used as stationary phase for GPC. The pore size of a gel must be carefully controlled in order to be able to apply the gel to a given separation. Other desirable properties of the gel forming agent are the absence of ionizing groups and, in a given solvent, low affinity for the substances to be separated. Commercial gels like PLgel, Sephadex, Bio-Gel (cross-linked polyacrylamide), agarose gel and Styragel are often used based on different separation requirements.

3.2.8.3. Eluent

The eluent (mobile phase) should be a good solvent for the polymer, should permit high detector response from the polymer and should wet the packing surface. The most common eluents in for polymers that dissolve at room temperature GPC are tetrahydrofuran (THF), o-dichlorobenzene and trichlorobenzene at 130–150°C for crystalline polyalkynes and m-cresol and o-chlorophenol at 90°C for crystalline condensation polymers such as polyamides and polyesters.

3.2.8.4. Pumps

There are two types of pumps available for uniform delivery of relatively small liquid volumes for GPC: piston or peristaltic pumps.

Low pressure peristaltic pumps are used for column chromatographic separation. They should have variable speed settings to control the flow rate of the column. An ideal pump should possess a positive or semi positive pumping action. They should be able to withstand back pressure up to 20 psi.

3.2.8.5. Detectors

In GPC, the concentration by weight of polymer in the eluting solvent may be monitored continuously with a detector. There are many detector types available and they can be divided into two main categories. The first is concentration sensitive detectors which includes UV absorption, differential refractometer (DRI) or refractive index (RI) detectors, infrared (IR) absorption and density detectors. Molecular weight sensitive detectors include low angle light scattering detectors (LALLS), multi angle light scattering (MALLS). Components collected on each tube are measured by direct spectrophotometric techniques. In sophisticated equipments, the outlet of the column is connected to an on-line detector, which monitors the components passed through a flow cell. These monitors give a continuous readout in the form of chromatogram that helps to collect the fractions of interest. The most popular detector is an absorbance spectrophotometer, which detects components by their ability to absorb light at defined wavelengths.

3.2.8.6. Fraction Collectors

Separated components are collected by placing a series of tubes under the outlet of the column and equal amounts of samples are collected in each tube. Sophisticated equipments use a fraction collector for the collection of samples from the column out flow automatically. The sample falls as drops into a collecting tube for a period of time, after which a new tube is mechanically moved into place. Fraction collectors are circular with the liquid dispensing arm traveling in a concentric path.

3.2.9. Technique

1. Gel preparation

Packing gel materials are swollen by soaking them in suitable buffer for a predetermined time prior to use. Swelling time varies with the type of gel materials. Space between the polymer chains increases during swelling. Swelling time can be reduced by heating the gel in boiling water bath for 1-5h.

2. Column packing

Swollen gel is degassed by applying vacuum and then packed into the column. Three volumes of running buffer is allowed to flow through the column.

3. Separation Technique

Conventional simple glass column are run under gravitational pressures in descending mode and it is time consuming. Desalting technique are performed under gravitational pressure. Modern technique use pumps to control the flow rate and achieve more reliable and reproducible separation of the sample components. Commercial columns with flow adapters are required to perform this separation.

3.2.10. Applications

One of the principal advantages of gel-filtration chromatography is that separation can be performed under conditions specifically designed to maintain the stability and activity of the molecule of interest without compromising resolution. Absence of a molecule - matrix binding step also prevents unnecessary damage to fragile molecules, ensuring that gel-filtration separations generally give high recoveries of activity.t

1. Separation of Proteins and Peptides

Because of its unique mode of separation, gel-filtration chromatography has been used successfully in the purification of literally thousands of proteins and peptides from various sources. These range from therapeutic proteins and peptides, which together constitute a multi-billion euro world-wide market, to enzymes and proteins for the brewing, food-processing and diagnostics industries

2. Separation of other Biomolecules

Carbohydrates represent a plentiful, but so far only scarcely exploited, reservoir of unique, multifunctional biopolymers which can be readily fractionated by gelfiltration chromatography on the basis of their relative sizes. Various problems, however, have limited the development of gel-filtration methods for oligosaccharides. Firstly, many of the commercially available gelfiltration matrices are themselves carbohydrates (*e.g.* Sephadex, Sepharose *etc.*, manufactured by GE Healthcare), shedding milligram quantities of heterodisperse carbohydrate polymers into the mobile phase. Secondly, non-specific interactions with matrix materials are common, since sugars are essentially amphipathic with a hydrophobic ring structure and hydrophilic functional groups. Despite these problems, however, gel-filtration chromatography still remains an important option for the purification of complex oligosaccharides.

3. *Separation of Cells and Virus Particles*

Cells of different sizes can be efficiently separated from one another using gelfiltration chromatography. Methods have been developed, for example, to separate both erythrocytes and platelets from blood (most workers now prefer density-gradient).

4. *Group Separations*

By selecting a matrix pore-size which completely excludes all of the larger molecules in a sample from the internal bead volume, but which allows very small molecules to enter this volume easily, one can effect a group separation in a single, rapid gelfiltration step which would traditionally require dialysis for up to 24 hours to achieve. Group separation can be used, for example, to effect buffer exchanges within samples, for desalting of labile samples prior to concentration and lyophilisation, to remove phenol from nucleic acid preparations, and to remove inhibitors from enzymes.

5. *Molecular Mass Estimation*

Gel-filtration chromatography is an excellent alternative to SDS-PAGE for the determination of relative molecular masses of proteins, since the elution volume of a globular protein is linearly related to the logarithm of its molecular weight.

6. *Size-Exclusion Reaction Chromatography: Protein PEGylation*

Covalent attachment of PEG (polyethylene glycol; "PEGylation") to a protein can attenuate its antigenicity and/or extend its biological half-life or shelf life. Size exclusion reaction chromatography (SERC) permits one to control the extent of a reaction (such as PEGylation) that alters molecular size and to separate reactants and products.

3.3. Ion Exchange Chromatography

3.3.1. Ion Exchange Chromatography

The most popular method for the purification of proteins and other charged molecules is ion exchange chromatography. Ion exchange chromatography (IEC) is a column chromatographic technique used for the separation and purification of proteins, peptides, nucleic acids, polynucleotides and other charged molecules with a high resolving power and high capacity.

3.3.2. Principle

The separation of molecules is achieved by passing the sample over an ion exchange matrix or ion exchanger. The ion exchangers are prepared by covalently linking charged groups or stationery phase to an insoluble support matrix. The support matrix are usually made of cellulose, dextran, agarose or silica. The separation of molecules in the sample is achieved in two stages.

- The molecule of interest gets adsorbed to the charged matrix while unbound material is removed
- The bound molecule is later recovered from the matrix by varying the pH or ionic strength of the buffer.

3.3.3. Types of Ion Exchangers

There are two types of ion exchanger, namely cation and anion exchangers.

- **Cation exchangers:** They have negatively charged groups and will attract positively charged cations. These exchangers are also called acidic ion exchangers because they attain negative charge due to the ionization of acidic groups.
- **Anion exchangers:** They have positively charged groups and will attract negatively charged anions. They are termed as basic ion exchangers, because they attain positive charges due to the association of protons with basic groups.

Mechanism

To optimize binding of all charged molecules, the mobile phase is generally a low to medium conductivity (*i.e.*, low to medium salt concentration) solution. The adsorption of the molecules to the solid support is driven by the ionic interaction between the oppositely charged ionic groups in the sample molecule and in the functional ligand on the support. The strength of the interaction is determined by the number and location of the charges on the molecule and on the functional group. By increasing the salt concentration (generally by using a linear salt gradient) the molecules with the weakest ionic interactions start to elute from the column first. Molecules that have a stronger ionic interaction require a higher salt concentration and elute later in the gradient. The binding capacities of ion exchange resins are generally quite high. This is of major importance in process scale chromatography, but is not critical for analytical scale separations.

Buffer pH

As a rule, the pH of the mobile phase buffer must be between the pI (isoelectric point) or pKa (acid dissociation constant) of the charged molecule and the pKa of the charged group on the solid support. For example, in cation exchange chromatography, using a functional group on the solid support with a pKa of 1.2, a sample molecule with a pI of 8.2 may be run in a mobile phase buffer of pH 6.0. In anion exchange chromatography a molecule with a pI of 6.8 may be run in a mobile phase buffer at pH 8.0 when the pKa of the solid support is 10.3.

Salt Gradients

As in most other modes of chromatography (SEC being the exception) a protein sample is injected onto the column under conditions where it will be strongly retained. A gradient of linearly increasing salt concentration is then applied to elute the sample components from the column. An alternative to using a linear gradient is to use a step gradient. This requires less complicated equipment and can be very effective to elute different fractions if the appropriate concentrations of salt are known, usually from linear gradient experiments.

Steps in an IEX Separation

An IEX medium comprises a matrix of spherical particles substituted with ionic groups that are negatively (cationic) or positively (anionic) charged. The matrix is usually porous to give a high internal surface area. The medium is packed into a column to form a packed bed. The bed is then equilibrated with buffer which fills the pores of the matrix and the space in between the particles. The pH and ionic strength of the equilibration buffer are selected to ensure that, when sample is loaded, proteins of interest bind to the medium and as many impurities as possible do not bind. The proteins which bind are effectively concentrated onto the column while proteins that do not have the correct surface charge pass through the column at the same speed as the flow of buffer, eluting during or just after sample application, depending on the total volume of sample being loaded. When the entire sample has been loaded and the column washed so that all non-binding proteins have passed through the column (*i.e.* the UV signal has returned to baseline), conditions are altered in order to elute the bound proteins. Most frequently, proteins are eluted by increasing the ionic strength (salt concentration) of the buffer or, occasionally, by changing the pH. As ionic strength increases, the salt ions (typically Na+ or Cl-) compete with the bound components for charges on the surface of the medium and one or more of the bound species begin to elute and move down the column. The proteins with the lowest net charge at the selected pH will be the first ones eluted from the column as ionic strength increases. Similarly, the proteins with the highest charge at a certain pH will be most strongly retained and will be eluted last. The higher the net charge of the protein, the higher the ionic strength that is needed for elution. By controlling changes in ionic strength using different forms of gradient, proteins are eluted differentially in a purified, concentrated form. A wash step in very high ionic strength buffer removes most tightly bound proteins at the end of an elution. The column is then re-equilibrated in start buffer before applying more sample in the next run.

3.3.4. Ion Exchanger Support Matrix

The support matrix are mainly made of cellulose, dextran, polyether, polystyrene, polyacrylate, agarose and silica. They have either an anionic or a cationic functional group derivatized to its surface. These two groups are further divided into weak and strong ion exchangers.

1. **Weak Ion exchanger:** Most common weak ion exchangers carry either the diethylaminoethyl (DEAE) or carboxymethyl (CM) groups derivatized to the support matrix. The former is an **anion exchanger** while the latter is a **cation exchanger**. Another weak anion exchanger is polyethyleneimine (PEI) which is used specifically to separate proteins and nucleic acids.
2. **Strong Ion exchanger:** Strong anion exchanger groups are quaternary aminoethyl (QAE), while strong cation exchangers carry sulfopropyl (SP) groups on their surface. These ion-exchange groups are fully ionized over a wide pH range.

Separation by adsorption of the molecule to the support matrix is preferable when several different molecules are to be separated. Separation by adsorption of the unwanted materials to the support matrix is preferable when single molecule is to be separated.

The different type of exchangers, their functional groups and the support matrix are given in the Table.

Type	*Functional Groups*	*Functional Group Name*	*Resins*
Weakly acidic	-COO-	Carboxy	Agarose
(cation exchanger)	$-CH_2COO-$	Carboxymethyl (CM)	Cellulose Dextran Polyacrylate
Strongly acidic	$-SO_3-$	Sulpho	Cellulose
(cation exchanger)	$-CH_2SO_3-$	Sulphomethyl	Dextran
	$-CH_2CH_2CHSO_3-$	Sulphopropyl	Polystrene, Polyacrylate
Weakly basic	$-CH_2CH_2N+H_3$	Aminoethyl	Agarose
(anion exchanger)	$-CH_2-CH_2N+H$	Diethylaminoethyl (DEAE)	Cellulose Dextran
	I		Polystyrene
	$(CH_2CH_3)_2$		Polyacrylate
Strongly basic	$-CH_2N+(CH_3)_3$	Trimethylaminomethyl	Cellulose
(anion exchanger)	$-CH_2CH_2N+(CH_2CH_3)$	Triethylaminoethyl	Dextran
	$-CH_2N+(CH_3)$	Dimethyl-2-hydroxyethyl-	Polystyrene
	I	aminomethyl	
	CH_2CH_2OH		

High porosity offers a large surface area covered by charged groups and so ensures a high binding capacity. High porosity is also an advantage when separating large biomolecules. Non-porous matrices are preferable for extremely high resolution separations when diffusion effects must be avoided. An inert matrix minimizes non-specific interactions with sample components. High physical stability ensures that the volume of the packed medium remains constant despite extreme changes in salt concentration or pH thus improving reproducibility and avoiding the need to repack columns. High physical stability and uniformity of particle size facilitate high flow rates, particularlyduring cleaning or re-equilibration steps, to improve through put and productivity. High chemical stability ensures that the matrix can be cleaned using stringent cleaning solutions if required.

Ion Exchange Matrices

Form	*Mean Particle Size*
Mini Beads Polystyrene/divinyl benzene	3 μm
Mono Beads Polystyrene/divinyl benzene	10 μm
SOURCE 15 Polystyrene/divinyl benzene	15 μm
SOURCE 30 Polystyrene/divinyl benzene	30 μm
Sepharose High Performance Agarose 6 per cent	34 μm
Sepharose Fast Flow Agarose 6 per cent	90 μm
Sepharose 4 Fast Flow Agarose 4 per cent	90 μm
Sepharose XL Agarose 6 per cent , dextran chains coupled to agarose	90 μm
Sepharose Big Beads Agarose 6 per cent	200 μm

Functional Groups Used on Ion Exchangers

Anion Exchangers	*Functional Group*
Quaternary ammonium (Q) strong	$-O-CH_2N+(CH_2)_3$
Diethylaminoethyl (DEAE)* weak	$-O-CH_2CH_2N+H(CH_2CH_3)_2$
Diethylaminopropyl (ANX)* weak	$-O-CH_2CHOHCH_2N+H(CH_2CH_3)_2$
Cation Exchangers	
Sulfopropyl (SP) strong	$-O-CH_2CHOHCH_2OCH_2CH_2CH_2SO_3-$
Methyl sulfonate (S) strong	$-O-CH_2CHOHCH_2OCH_2CHOHCH_2SO_3-$
Carboxymethyl (CM) weak	$-O-CH_2COO-$

3.3.4.1. Polystyrene

Polystyrene resins are made by co-polymerising styrene with the cross linker, divinyl benzene. Sulphonation of cross-linked polystyrene gives sulphonated polystyrene resin which is a strong cation exchanger (*e.g.* Dowex 50). Anion exchangers are prepared by reacting cross-linked polystyrene with chlormethylether and then the chlorogroup is ionized with tertiary amine $-CH_2N^+(CH_3)_3Cl^-$ groups.

3.3.4.2. Cellulose

Cellulose is a high molecular weight compound. Carboxymethyl cellulose (CM-cellulose) is formed when the $-CH_2OH$ group is converted to $-CH_2OCH_2COOH$, which of sample, it can be used as a preparative method. For the latter, the separated substances are recovered by scraping the adsorbent off the plate (or cutting out the spots if the supporting material can be cut) and extracting the substance from the adsorbent. Several factors determine the efficiency of a chromatographic separation. The adsorbent should show a maximum of selectivity toward the substances being separated so that the differences in rate of elution will be large. For the separation of any given mixture, some adsorbents may be too strongly adsorbing or too weakly adsorbing. Table lists a number of adsorbents in order of adsorptive power. Silica gel is the most common adsorbent used for routine TLC of organic compounds.

1. The different compounds in the sample mixture migrate at different rates due to the differences in their attraction to the stationary phase, and differences in solubility in the mobile phase. The separation of compounds is therefore based on the competition of the compound and mobile phase for the binding sites on the stationary phase.
2. Silica gel is used as the stationary phase is polar. So, the more polar compound has a stronger interaction with the silica gel and it dispels the mobile phase from the binding sites. Consequently, the less polar compound moves higher up the plate. If the mobile phase is changed to a more polar solvent or solvent mixture, it dispels the compounds from the binding sites and all compounds on the plate will move higher up the plate. Strong solvents move the compounds up the plate, while weak solvents do not move them.
3. The order of strength/weakness depends on the stationary phase. In silica gel coated TLC plates, the solvent strength increases in the following order: perfluoroalkane, hexane, pentane, carbon tetrachloride, benzene/ toluene, dichloromethane, diethyl ether, ethylacetate, acetonitrile, acetone, 2-propanol/n-butanol, water, methanol, triethylamine, acetic acid, formic acid. In C-18 coated TLC plates, the order is reverse. An eluotropic series is mainly used as a guide in the selection of the mobile phase.

3.5.2. Adsorbents

The common adsorbents used in TLC are silica, alumina, kieselguhr, celite, cellulose, starch and Sephadex. Table gives the list of adsorbents used in TLC and their applications in the separation of different compounds.

The order in the table is approximate, since it depends upon the substance being adsorbed and the solvent used for elution.

Most Strongly Adsorbent

- ☆ Alumina Al_2O_3
- ☆ Charcoal C
- ☆ Florisil MgO/SiO_2 (anhydrous)

is a weak cation exchanger. Similarly, DEAE-cellulose is formed when the $-CH_2OH$ is converted to $CH_2OCH_2CH_2N(CH_2CH_3)_2$, which is a weak anion exchanger. These are the popular exchangers used for the separation of proteins.

3.3.4.3. Sepharose and Sephadex

Sepharose resins are derived from cross-linked agarose, while Sephadex resins are derived from cross-linked dextrans. These resins are used for the separation of high molecular weight proteins and nucleic acids.

3.3.4.4. Choice of Ion Exchanger

The selection of an ion exchanger is based on molecule stability, molecular weight and effect of pH on charge. Many biological molecules, especially proteins are stable within only a fairly narrow pH range. The ion exchanger selected must therefore operate within this range. The nature of resins, their optimum pH range and applications are given in the Table.

Nature of Resin	*pH Range*	*Applications*
Anion exchangers		
Strong	2 - 11	Nucleotides
Intermediate	2 - 7	Organic acids
Weak	3 - 6	Proteins
Cation exchangers		
Strong	2 - 11	Amino acids
Intermediate	6 - 10	Peptides
Weak	7 - 10	Proteins

If the molecule is stable over a wide range of pH, either an anion or a cation exchanger can be used. For example, the weak electrolytes that require very low or high pH for ionization are separated only by strong exchangers, as they operate over a wide pH range. In case of the strong electrolytes, weak exchangers are advantageous because they reduce protein denaturation, inability to bind weakly charged impurities and enhance elution characteristics.

3.3.5. Eluent Buffers

There are anionic and cationic buffers for the elution of compounds.

- ✰ Cationic buffers are tris, pyridine and alkyl amines. They are used with anion exchangers.
- ✰ Anionic buffers are acetate, barbiturate and phosphate. They are used with cationic exchangers.

The pH of the buffer selected as eluent should be one pH unit above or below the isotonic point of the molecule. The initial pH and ionic strength should allow the binding of the molecules to the exchanger. A buffer of the lowest ionic strength should be initially used as it ensures that minimum binding of contaminants to the exchanger. If gradient elution is used, initial conditions should allow all the sample compounds to bind at the top of the column.

3.3.6. Elution

Gradient elution is most common than isocratic elution. In isocratic elution, the sample volume is 1-5 per cent of bed volume and in gradient elution, large volumes can be applied. Continuous or stepwise pH and ionic strength gradients can be employed but a continuous gradient tends to give better resolution. During the elution with an anion exchanger, pH gradient decreases and ionic strength increases, whereas with a cation exchanger, both the pH and ionic gradients increase.

3.3.7. Technique

- The sample is applied on top of the column which contains the ion exchange resin.
- The molecules of interest get bound or adsorbed to the column.
- The column is washed with the equilibration buffer of low ionic strength.
- The impurities or unwanted ions that are not bound are eluted off using this buffer.
- The bound molecules are then eluted using a gradient buffer which steadily increases the ionic strength of the eluent.
- Alternatively, the pH of the eluent buffer can also be modified so as to give the molecule or the resin matrix a charge at which they will not interact.
- Finally, the molecule of interest is eluted from the resin.

3.3.8. Applications

- Separation of amino acids by strong acid cation exchanger such as Dowex 50-PS and SP-Sephadex-cellulose
- Separation of proteins by weakly acidic and basic exchangers derived from cellulose and agarose
- Determination of base composition of nucleic acids.

3.4. Affinity Chromatography

Affinity chromatography is a form of adsorption chromatography in which separation and purification of molecules are achieved on the basis of their specific biological interactions. It is separating biochemical mixtures based on a highly specific interaction such as that between antigen and antibody, enzyme and substrate, or receptor and ligand. Affinity chromatography is a method of separating biochemical mixtures based on a highly specific interaction such as that between antigen and antibody, enzyme and substrate, or receptor and ligand.

This technique was originally developed for the purification of enzymes and later extended to nucleotides, nucleic acids, immunoglobins, membrane receptors, whole cells and cell fragments. This technique has got high selectivity and high resolution. The molecules are purified in concentrated form several thousand-fold with high recovery by this technique.

3.4.1. Principle

- ☆ Separation and purification of a molecule is based on a reversible interaction between the molecule and a specific ligand that is immobilized to a insoluble chromatography matrix.
- ☆ Molecule (M) + Ligand (L) = ML complex
- ☆ A sample containing a specific molecule of interest is passed through a column containing immobilized ligand. The specific molecule will then bind to the ligand to form a ML complex. This biological interaction between ligand and specific molecule occurs due to electrostatic or hydrophobic interactions, van der Waals' forces or hydrogen bonding. The molecules that are not specifically bound get washed away with the buffer. The molecules that are specifically bound are recovered from the ligand by reversing the interaction.

The stationary phase is typically a gel matrix, often of agarose; a linear sugar molecule derived from algae. Usually the starting point is an undefined heterogeneous group of molecules in solution, such as a cell lysate, growth medium or blood serum. The molecule of interest will have a well known and defined property, and can be exploited during the affinity purification process. The process itself can be thought of as an entrapment, with the target molecule becoming trapped on a solid or stationary phase or medium. The other molecules in the mobile phase will not become trapped as they do not possess this property. The stationary phase can then be removed from the mixture, washed and the target molecule released from the entrapment in a process known as elution. Possibly the most common use of affinity chromatography is for the purification of recombinant proteins.

Binding to the solid phase may be achieved by column chromatography whereby the solid medium is packed onto a column, the initial mixture run through the column to allow setting, a wash buffer run through the column and the elution buffer subsequently applied to the column and collected. These steps are usually done at ambient pressure. Alternatively, binding may be achieved using a batch treatment, for example, by adding the initial mixture to the solid phase in a vessel, mixing, separating the solid phase, removing the liquid phase, washing, re-centrifuging, adding the elution buffer, re-centrifuging and removing the eluate.

Components of Affinity Chromatography

- ☆ **Matrix:** for ligand attachment. Matrix should bechemically and physically inert.

- **Spacer arm:** used to improve binding between ligand and target molecules by overcoming any effects of steric hindrance.
- **Ligand:** molecule that binds reversibly to a specific target molecule or group of target molecules.
- **Binding:** buffer conditions are optimized to ensure that the target molecules interact effectively with the ligand and are retained by the affinity medium as all other molecules wash through the column.
- **Elution:** buffer conditions are changed to reverse (weaken) the interaction between the target molecules and the ligand so that the target molecules can be eluted from the column.
- **Wash:** buffer conditions that wash unbound substances from the column without eluting the target molecules or that re-equilibrate the column back to the starting conditions(in most cases the binding buffer is used as a wash buffer).
- **Ligand coupling:** covalent attachment of a ligand to a suitable pre-activated matrix to create an affinity medium.
- **Pre-activated matrices:** matrices which have been chemically modified to facilitate the coupling of specific types of ligand.

3.4.2. Affinity Chromatography Support Matrices

The support matrices should have large surface area and suitable side chains for ligand attachment. They should be chemically inert to avoid nonspecific adsorption during the separation. There are many types of support matrices.

- Organic supports
- Inorganic supports
- Synthetic supports
- Composite supports.

3.4.2.1.Organic Supports

They are supports made of biological materials such as cellulose, dextran and agarose.

- **Cellulose:** Cellulose was the first material used for attachment of affinity ligands mainly for the isolation of enzymes. They do not promote fast flow rates, but are chemically inert and cheap.
- **Dextran:** Dextran is a microbial polysaccharide. They are cross-linked with epichlorohydrin to produce a beaded gel. They are highly hydrophilic due to OH groups, which are used to attach ligand.
- **Agarose:** Agarose is the most popular material used as an affinity support. They produce hydrophilic beads. The major disadvantage is that they have a tendency to non specifically bind proteins and nucleic acids. *e.g.* Sepharose and Superose 6B

3.4.2.2. Inorganic Supports

Inorganic material used as support matrices include silica and glass. They are mainly used for the immobilization of enzymes.

- **Silica:** Silica is chemically modified to produce silane groups for attachment of side chains such as epoxy or thio groups. These side chains provide attachment for the ligand.
- **Solid Glass Beads:** Solid glass beads are not available with reactive side chains. They are first silanized and modified to attach the reactive side chains. They have high flow rates and used for construction of preparative columns.

3.4.2.3. Synthetic Supports

Synthetic support matrixes are polymers of plastics such as polyacrylamide, methacrylate and acrylic. They possess high mechanical strength and maintain high flow rates.

- **Polyacrylamide:** Polyacrylamide beads contain a hydrocarbon structural framework with carboxamide side chains, which are chemically modified to acyl groups and used to attach protein ligands to the beads.
- **Methacrylate:** In methacrylate beads, ligand is coupled through OH groups present on the surface of the beads. These are used for hydrophobic interaction separations of proteins and peptides.
- **Acrylic:** In oxirane acrylic beads, ligand immobilization is done through activation of oxirane groups on the bead matrix.

3.4.2.4. Composite Supports

There are two types of composite supports such as polysaccharide - polyacrylamide and magnetic beads.

- **Polysaccharide-Polyacrylamide:** Combination of polysaccharides with acrylamide polymers provides a support that possesses the tensile strength of polyacrylamide and the high molecular weight sieving qualities of the polysaccharide gels. The amide groups of the polyacrylamide and OH groups of allyl dextran are used for ligand attachment. *e.g.* Sephacryl HR.
- **Magnetic beads:** Magnetic beads are used as ligand supports for batch separation, in which the affinity ligand is recovered with a magnet. *e.g.* Magnogel

3.4.3. Types of Ligands

- There are different molecules that can be used as specific ligands based on their specific or reversible binding properties. Important ligands include proteins, nucleic acids, lectins and textile dyes.

☆ **Proteins:** Protein ligands include immobilized enzymes, coenzymes, antibodies, antigens, proteins and hormones. Immobilized enzymes are used to isolate both coenzymes and substances. Proteins ligands such as Protein A, Protein G, avidin and streptavidin are used to isolate monoclonal antibodies from cell cultures and for the isolation of biotin-labeled molecules. Immobilized hormones are used to isolate specific receptors as peptide antigens.

☆ **Nucleic Acids:** DNA and RNA are used as immobilized ligands for the isolation of proteins which specifically bind to nucleic acids.

☆ **Lectins:** Lectin, a plant protein has strong affinity to bind certain carbohydrate residues. *e.g.* Concanavalin A

☆ **Textile Dyes:** Synthetic textile dyes with a wide variety of biological materials are used as affinity ligands. They are suitable for the isolation of lipoproteins, α-fetoprotein, IgG, α_2-macroglobulin and human clotting factors. *e.g.* Cibacron Blue

Table gives the types of ligands and their specific affintity towards different molecules.

Ligand	*Molecules*
5'AMP	NAD+ dependent dehydrogenases, some kinases
2'5'-ADP	NADP+ dependent dehydrogenases
Calmodulin	Calmodulin binding enzymes
Avidin	Biotin containing enzymes
Fatty acids	Fatty acids binding proteins
Heparin	Lipoproteins, lipases, coagulation factors, DNA polymerases, steroid receptor proteins, growth factors, serine protease inhibitors
Proteins A and G	Immunoglobulins
Concanavalin A	Glycoproteins containing a -D-mannopyranosyl and a -D-glucopyranosyl residues
Soybean lectin	Glycoproteins containing N-acetyl a -(or b) -D-galactopyranosyl residues
Phenylboronate	Glycoproteins
Poly (A)	RNA containing poly (U) sequences, some RNA specific proteins
Lysine	rRNA
Cibacron Blue FF3G-A	Nucleotide-requiring enzymes, coagulation factors

3.4.4. Immobilization Methods

☆ Ligands are coupled to the support matrix by a series of reactions. The reactive side chains of the support matrix are chemically modified before the attachment of the ligand by different activation methods such as cyanogen bromide, triazine, periodate oxidation, epoxide, acyl azide, diazo, succinate esters, carbodiimide and carbonylation.

- ☆ All the reactive side chains in the support matrix are not coupled to the ligand. There are uncoupled side chains that can bind with specific and non-specific components. Blocking agents are used to block the uncoupled side chains. *e.g.* Human IgG is used as a blocking agent to isolate human IgE.

3.4.5. Spacer Arms

After the activation of the support matrix, a suitable spacer molecule is attached between the support and the ligand. Spacer molecule helps to prevent the hindrance between the immobilized support and the molecules to be isolated. They are important for small immobilized ligands such as an antigen and not for macromolecular ligands. Spacer arm ensures that the ligand is placed at a suitable distance from the surface of the support. Spacer arms are either attached directly to support surface or to the reactive side chain of the support matrix. Optimum length of the spacer arm is six to ten carbon atoms. Some spacer arms are purely hydrophobic consisting of methylene (CH_2) groups and others are hydrophilic consisting of carbonyl (CO) or imido (NH) groups. Several matrix supports of agarose, dextran and polyacrylamide are available with a variety of spacer arms and ligands attached for immediate use.

3.4.2. Elution Techniques

The ligand bound molecule can be recovered by different elution techniques. Elution technique involves mainly competitive elution, in which the ligand bound molecule is recovered in the presence of a competing agent.

Elution Agents

- ☆ **Ligand competition:** Addition of free ligand to displace the adsorbed molecule forms the immobilized matrix. This is a highly specific technique but highly expensive. Use of co-substrate has therefore proven to be the most efficient and inexpensive type of elution.
- ☆ **pH change:** pH manipulation is the most popular technique for dissociation of ligand bound molecule. The pH of the running buffer are changed to acid range of 1-1.5 in a gradient manner using citric, formic and acetic acids. Silica support is sensitive to sharp pH changes. Alkaline treatment is used for low pressure affinity but has greater denaturing effects.
- ☆ **Chaotropic ion elution**: There are several salts that have chaotropic characteristics. They are CCl_3COO^-, SCN^-, CF_3COO^-, ClO_4^-, I^-, Cl^- and used in concentrations between 1.5 and 8 M. The most widely used chaotropic ion is sodium thiocyanate at 3 M concentrations. They cause some denaturation to protein ligands and substrates.

Elution Techniques

There are two basic techniques for recovering the molecules bound to ligands. They are step gradient and linear gradient elution.

- **Step Gradient Elution:** In step gradient elution, the recovery of bound molecule is achieved by a series of changes. It is easy to perform but causes loss of activity in the recovered molecule.
- **Linear Gradient Elution:** This is the most efficient method for the recovery of bound molecules from the ligands. Gradient making instrument that introduce the elution buffer in a linear concentration gradient is used.

Another classification of elution technique is specific and non-specific elution.

- **Non-specific elution:** Change of pH or ionic strength or polarity of the elution buffer recovers the bound molecules form the ligands. Change in pH elution using dilute acetic acid or ammonium hydroxide causes elution due to the ionization of groups in ligand and molecule. Change in ionic strength with or without concomitant change in pH causes elution due to a disruption of the ligand-molecule interaction.
- **Specific elution:** Specific clution is achieved by the use of a soluble competitive ligand. A preliminary knowledge on the structure and biological specificity of the molecule is required to choose the competitive ligand. *e.g.* In case of an enzyme, the ligand can be a competitive reversible inhibitor or an allosteric modifier.

3.4.8. Applications

The application of this technique is limited based to the availability of immobilized ligand. Some of them include

- Purification of many enzymes, proteins, receptor proteins and immunoglobulins.
- Isolation of messenger RNA by selective hybridization of poly U – Sepharose 4B by exploiting its poly (U) tail.
- Isolation of complementary RNA and DNA by immobilized single stranded DNA
- Isolation of proteins involved in nucleic acid metabolism by immobilized nucleotides.

3.5. Thin Layer Chromatography (TLC)

Thin layer chromatography (TLC) is a useful technique for the separation and identification of compounds in mixtures. TLC is used routinely to follow the progress of reactions by monitoring the consumption of starting materials and the appearance of products. Commercial applications of TLC include the analysis of urine for evidence of "doping", the analysis of drugs to establish purity or identity of the components, and analysis of foods to determine the presence of contaminants such as pesticides. **Thin layer chromatography** (TLC) is a chromatography technique used for the separation of the compounds from a sample mixture. It is performed on a sheet of glass, plastic, or aluminum foil coated with a thin layer of adsorbent known as the stationery phase. Solvent or solvent mixture is used as the "mobile phase", which is also known as developing solvent. Mobile phase moves the sample applied on the plate through the stationery phase by capillary action and effects the separation of compounds. This technique was invented by Schraiber in 1939.

3.5.1. Principle

Thin layer chromatography (TLC) uses the same principles as extraction to accomplish the separation and purification of compounds: that is, the different separation of compounds between two phases based on differences in solubility of compounds in the two phases. In the case of TLC, one phase is a mobile liquid solvent phase and the other phase is a stationary solid phase with a high surface area. The stationary phase normally consists of a finely divided adsorbent, silica (SiO_2) or alumina (Al_2O_3) powder, used in the form of a thin layer (about 0.25 mm thick) on a supporting material. The support is usually a sheet of glass or metal foil. The mobile phase consists of a volatile organic solvent or mixture of solvents. A solution of the sample containing a mixture of compounds is applied to the layer of adsorbent, near one edge, as a small spot. The TLC plate is propped vertically in a closed container (developing chamber), with the edge to which the spot was applied down. The solvent, which is in the bottom of the container, travels up the layer of adsorbent by capillary action, passes over the spot and, as it continues up, moves the compounds in the mixture up the plate at different rates resulting in separation of the compounds. This process of moving the compounds with the solvent is referred to as elution and the solvents used are eluting solvents. This overall procedure is referred to as "developing" the TLC plate. When the solvent front has nearly reached the top of the stationary phase, the plate is removed from the container, and the solvent front is marked with a pencil.

In the diagram to the right, a single spot from a reaction mixture reveals that there are 2 components (A and B) in that mixture.

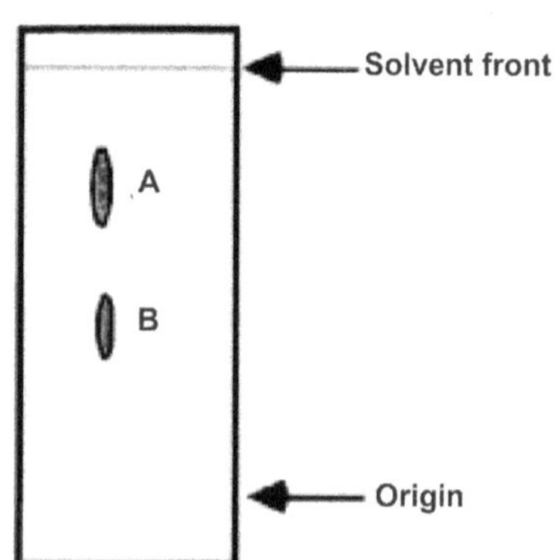

Since the amount of adsorbent involved is relatively small, and the ratio of adsorbent to sample must be high, the amount of sample must be very small, usually less than a milligram. For this reason, TLC is often used as an analytical technique rather than a preparative method, although with thicker layers (about 2 mm) and large plates with a number of spots or a stripe

Adsorbents and their Application

Adsorbent	*Applications*
Silica	Amino acids, alkaloids, sugars, fatty acids, lipids, essential oils, inorganic anions and cations, steroids, terpenoids.
Alumina	Alkaloids, food dyes, phenols, steroids, vitamins, carotenes, amino acids.
Kieselguhr	Sugars, oligosaccharides, dibasic acids, fatty acids, triglycerides, amino acids and steroids.
Celite	Steroids.
Cellulose powder	Amino acids, food dyes, alkaloids, nucleotides.
Ion exchange cellulose	Nucleotides.
Starch	Amino acids.
Sephadex	Amino acids, proteins.

Least Strongly Adsorbent

- ☆ Silica gel SiO_2

The eluting solvent should also show good selectivity in its ability to dissolve or desorb the substances being separated. The solubility of different compounds in the eluting solvent plays an important role in how fast they move up the TLC plate. However, a more important property of the solvent is its ability to itself be adsorbed on the adsorbent. To the extent that the solvent has affinity for the adsorbent, it can displace the compounds being separated thereby "pushing" them up the plate. If the solvent is too strongly adsorbed, it can fully displace all compounds causing them to move up the plate together near the solvent front with no separation. If the solvent is too weakly adsorbed, its solvating power alone may be insufficient to move any compounds fast enough to effect separation. Ideally, the affinity of the eluting solvent for the adsorbent is comparable to the compounds being separated causing different compounds to move at different rates resulting in separation.

Lists of a number of common solvents in order of increasing eluting strength given below. The eluting strength of a solvent is primarily related to how strongly it adsorbs onto the adsorbent and because typical adsorbents are highly polar; thus, eluting strength increases with solvent polarity.

Solvents for Chromatography

- ☆ Pentane, hexane, heptane
- ☆ Toluene, p-xylene
- ☆ Dichloromethane
- ☆ Diethyl ether (anhydrous)
- ☆ Ethyl acetate (anhydrous)
- ☆ Acetone (anhydrous)
- ☆ Acetic acid

☆ Ethanol (anhydrous)

☆ Methanol (anhydrous)

Calculating the Rf Value of a Compound

The distance travelled by a compound relative to the distance travelled by the solvent front depends upon the structure of the molecule, and so TLC can be used to identify compounds as well as to separate them. The relationship between the distance traveled by the solvent front and the compound is usually expressed as the Rf value:

Rf value = distance travelled by compound/distance travelled by solvent front

Rf values are strongly dependent upon the nature of the adsorbent and solvent system and thus experimental Rf values and literature values do not always agree. In order to determine whether an unknown compound is identical to a compound of known structure, it is necessary to run the two samples side by side on the same TLC plate, preferably at the same concentration. In general, **low polarity compounds have higher Rf values than higher polarity compounds.**

3.5.3. Plate Preparation

TLC plates are prepared by mixing the adsorbent *e.g.* silica gel with a small amount of inert binder such as calcium sulfate or gypsum and water. This mixture is spread as thick slurry on an unreactive carrier sheet, such as glass or thick aluminum foil or plastic. The resultant plate is dried and activated by heating in an oven for 30 min at 110°C. The thickness of the adsorbent layer is typically around 0.1 – 0.25 mm for analytical purposes and 0.5 – 2.0 mm for preparative purposes. Precoated TLC plates are commercially available with standard particle sizes with improved reproducibility.

3.5.4. Technique

How to Run a TLC Plate

1. Draw a pencil line about a 1/4 inch from the bottom of the plate (along the short side). Mark places along the line for each spot (pure reference spots and your mixture).
2. Dip the capillary into the solution and gently and quickly place a 1-2 millimetre spot on the plateat the position you've marked. Keep the spots small!
3. Pour approx. 3 ml of solvent into a screw-cap jar, place a piece of filter paper in the jar and wet the paper with the solvent to saturate the atmosphere. Make sure that the solvent is shallow enough that it will be below the spot line on your plate.
4. Place the plate carefully in the chamber (use tongs or tweezers), being careful not to dunk thespots under the solvent. Put the cap on the jar and let the plate develop.

5. When the solvent front has almost reached the top of the TLC plate, remove the plate and immediately mark the solvent front with a pencil line.
6. Using tweezers, dip the TLC plate into the solution of PAA stain in a fume hood. The plate should be dipped smoothly in one motion – your TA will demonstrate.
7. Wipe excess stain from the back of the plate (glass part) and place it on a hot plate for ~20 sec. allow the solvent to evaporate and look at the spots.
8. Circle the spots in pencil, bre ak out a ruler and calculate Rf values.

Factors to be Considered for TLC Experiment

- **Sample application:** Sample dissolved in a solvent is applied by means of micropipette or syringe as spots. Spot is placed 2-2.5 cm from the edge of plate. Solvent in the sample can be removed by gentle heating or by use of air blower. If a non-volatile solvent is used, the plate should be dried in a vacuum chamber.
- **Plate development:** The separation takes place in a TLC glass chamber which contains the developing solvent or mobile phase to a depth of about 1.5 cm. The solvent is allowed to stand for at least 1 h with top cover plate to saturate the atmosphere within the chamber with solvent vapour. Failure to saturate the chamber will result in poor separation and non- reproducible results. Then, the plate is placed vertically in the tank and the cover plate is replaced. The separation occurs as the solvent moves up the plate by capillary action. When the solvent front reaches the top of the plate in the chamber, the plate is removed and air-dried. The speed of separation is 10-30 min commonly and > 90 min occasionally.
- **Detection:** Compounds separated on TLC are mostly colourless and there are several detection methods to visualize the colourless spots.
  - Spray of 50 per cent H_2SO_4 or 25 per cent H_2SO_4 in ethanol followed by application of heat to visualize as brown spots.
  - Addition of a fluorescent compound such as manganese - activated zinc silicate to the adsorbent to visualize the spots as fluorescent green under a UV 254nm light.
  - Subject the plates to iodine vapour to identify the unsaturated compounds.
  - Spray of specific color reagents to identify certain compounds *e.g.* ninhydrin for aminoacids.
  - Subject the plates to autoradiography to detect spots as dark areas on X ray film.
  - Scan the plates by radio chromatogram scanner, if the compounds are radioactive.

3.5.5. Retention Factor

Once the spots of compounds are visible, the "Rf value" or "retention factor" of each spot can be determined by dividing the distance travelled by the compound by the total distance travelled by the solvent.

- ☆ Rf value = {Distance moved by solute or compound over Distance moved by solvent}
- ☆ This value is constant for a particular compound under standard conditions. The compounds are then identified based on reference compounds separated alongside the sample.

3.5.6. Quantification

On-plate quantification is achieved by the use of radio chromatogram scanner for radio-labeled compounds or by the UV or visible absorption of compound in the densitometer. Off-plate quantification is carried out by scraping off the spot and eluting the compound with suitable solvent. The amount of compound in solution is then determined by spectrophotometry.

3.5.7. Preparative TLC

Preparative TLC is used on semi-preparative scale to separate compounds upto a few hundred milligram. In this technique, the sample mixture is not "spotted" on the TLC plate as dots, but is applied as a thin even layer horizontally to and just above the solvent level as a band. The compounds separate as horizontal bands rather than as spots upon development with the solvent. Each band is then scraped off, extracted with a suitable solvent and filtered to get the preferred compound upon removal of the solvent. Preparative TLC is far more efficient in terms of time and cost than column chromatography. This technique is best used for compounds that are coloured or visible under UV light.

3.5.8. Two Dimensional TLC

Two-dimensional thin layer chromatography is performed to improve resolution of a particular separation. In this technique, the material is placed towards one corner of plate as single spot and developed in one direction and allowed to dry. Then, it is again developed by another solvent system in the direction at right angles to first development.

3.5.9. Applications

Thin layer chromatography is used to identify compounds present in a given mixture, to determine the purity of a substance and to monitor the progress of a reaction. Some of the specific applications include:

1. Determination of the plant components
2. Analysis of cereamides and fatty acids

3. Detection of pesticides or insecticides in food and water
4. Analysis of the dye composition of fibers in forensics
5. Analysis of the radiochemical purity of radiopharmaceuticals

3.6. Gas Chromatography

Gas solid chromatography (GSC) was first invented by German Fritz Prior in 1947. Gas liquid chromatography (GLC) was later invented by Archer John Porter Martin in 1950 and it is now popularly known as Gas Chromatography. Gas chromatography - specifically gas-liquid chromatography - involves a sample being vaporized and injected onto the head of the chromatographic column.

3.6.1. Two Techniques

Gas-Solid Chromatography (GSC)

Gas solid chromatography has a solid stationary phase that physically adsorbs the compounds leading to their retention on the column. It is mainly applicable for the separation of specific low molecular weight compounds. It retains active or polar molecules for longer time and cause severe tailing of elution peaks that have restricted their use.

Gas-Liquid Chromatography (GLC)

Gas liquid chromatography has a thin layer of a liquid stationary phase immobilised inside the column and the compounds get partitioned between gas mobile phase and liquid stationery phase. It is the widely adopted and the most useful technique.

3.6.2. Principle

The organic compounds are made volatile by derivatization prior to their separation in gas chromatography. Very small amounts of derivatized liquid compound mixture are injected into GC, gets volatilized in the hot injection chamber. The inert carrier gas used as a mobile phase carries the volatized compound mixture through a heated column. The column contains a high boiling liquid stationary phase adsorbed onto the surface of an inert solid material. The compounds that flow through the column get separated based on their differences in their partitioning behaviour between mobile gas phase and stationary phase in the column. The separated compounds then pass through a detector that detects the individual compounds and sends an electronic message to the recorder. The recorder responds to the message and prints the chromatogram.

3.6.3. Instrumentation

Schematic diagram of a gas chromatograph:

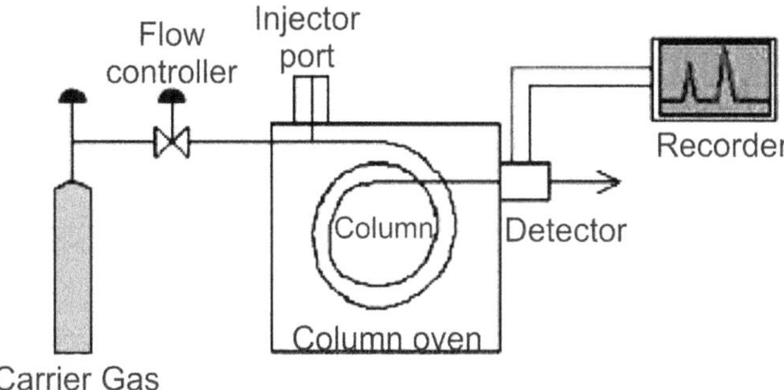

Instrumental Components

3.6.3.1. Carrier Gas

The carrier gas must be chemically inert. Commonly used gases include nitrogen, helium, argon, and carbon dioxide. The choice of carrier gas is often dependent upon the type of detector which is used. The carrier gas system also contains a molecular sieve to remove water and other impurities. They should have a constant flow rate. The faster the flow rate, the lower will be the retention time. The flow rate of the carrier gas is generally set at 55 ml/min. The choice of carrier gas is dependent upon the type of detector. *e.g.* FID uses helium gas.

3.6.3.2. Injector

Injector port contains a heated chamber into which sample is injected through a rubber septum using micro syringe. Septum purge outlet in the injector port prevents septum bleed components from entering into the column. It is always good to introduce as small a volume for optimum column efficiency. If it is a packed column, inject 1 to 5 μl and a capillary column, inject 0.5 μl volumes. The sample vaporizes in the injector port to form a mixture of carrier gas and vaporized solutes. It is important to maintain the injector at a temperature of about 50°C higher than the boiling point of the least volatile component in the sample mixture. There are two types of injection known as split injection and split less injection. In split injection mode, a portion of the sample compound mixture passes into the column and rest exit through the split outlet.

Sample Injection Port

For optimum column efficiency, the sample should not be too large, and should be introduced onto the column as a "plug" of vapour - slow injection of large samples causes band broadening and loss of resolution. The most common injection method is where a micro syringe is used to inject sample through a rubber septum into a flash vaporiser port at the head of the column. The temperature of the sample port is usually about 50°C higher than the boiling point of the least volatile component of the sample. For packed columns, sample size ranges from tenths of a micro litre up to 20 micro litres. Capillary columns, on the other hand, need much less sample,

Gas Recommendation for Capillary Columns

Detector	Carrier gas	Preferred makeup gas	Second choice	Detector, anode purge, or reference gas
Electron Capture	Hydrogen	Argon/Methane	Nitrogen	Anode purge must be same as makeup
	Helium	Argon/Methane	Nitrogen	
	Nitrogen	Nitrogen	Argon/Methane	
	Argon/Methane	Argon/Methane	Nitrogen	
Flame Ionization	Hydrogen	Nitrogen	Helium	Hydrogen and air for detector
	Helium	Nitrogen	Helium	
	Nitrogen	Nitrogen	Helium	
Flame Photometric	Hydrogen	Nitrogen		Hydrogen and air for detector
	Helium	Nitrogen		
	Nitrogen	Nitrogen		
	Argon	Nitrogen		
Nitrogen Phosphorus	Helium	Nitrogen	Helium**	Hydrogen and air for detector
	Nitrogen	Nitrogen	Helium**	
Thermal Conductivity	Hydrogen*	Must be same as carrier and reference gas	Mus t be same as carrier and reference gas	Reference must be same as carrier and makeup
	Helium			
	Nitrogen			

* When using hydrogen with a thermal conductivity detector, vent the detector exhaust to a fume hood or a dedicated exhaust to avoid buildup of hydrogen gas.
** Helium is not recommended as a makeup gas at flow rates> 5ml/min. Flow rates above 5ml/min shorten detector life.

Gas Recommendation for Packed Columns

Detector	Carrier gas	Comments	Detector, anode purge, or reference gas
Electron Capture	Nitrogen	Maximum sensitivity	Nitrogen
	Argon/Methane	Maximum dynamic range	Argon/Methane
Flame Ionization	Nitrogen	Maximum sensitivity	Hydrogen and air for detector
	Helium	Acceptable alternative	
Flame Photometric	Hydrogen		Hydrogen and air for detector
	Helium		
	Nitrogen		
	Argon		
Nitrogen Phosphorus	Helium	Optimum performance	Hydrogen and air for detector
	Nitrogen	Acceptable alternative	
Thermal Conductivity	Helium	General use	Reference must be same as carrier
	Hydrogen	Maximum sensitivity (Note A)	
	Nitrogen	Hydrogen detection (Note B	
	Argon	Maximum hydrogen sensitivity (Note B)	

Note A: Slightly greater sensitivity than helium. Incompatible with some compounds.
Note B: For analysis of hydrogen or helium. Greatly reduces sensitivity for other compounds.

typically around 10-3 μL. For capillary GC, split/split less injection is used. Have a look at this diagram of a split/split less injector.

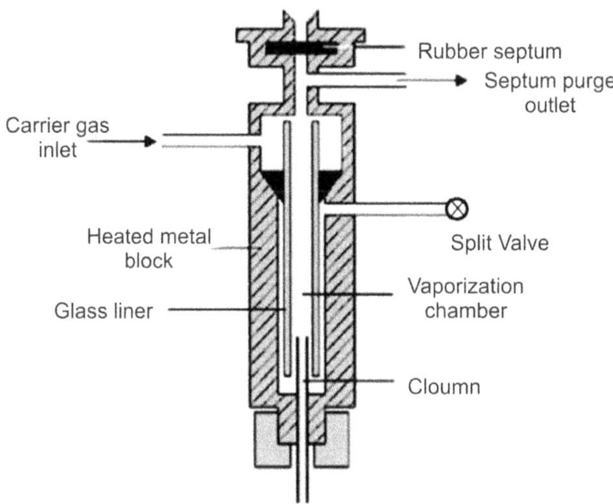

The injector can be used in one of two modes; split or splitless. The injector contains a heated chamber containing a glass liner into which the sample is injected through the septum. The carrier gas enters the chamber and can leave by three routes (when the injector is in split mode). The sample vapourises to form a mixture of carrier gas, vapourised solvent and vapourised solutes. A proportion of this mixture passes onto the column, but most exits through the split outlet. The septum purge outlet prevents septum bleed components from entering the column.

3.6.3.3. Column

The separation of components in the mixture takes place in the column on the basis of their interaction with stationary and mobile phases. The separation depends on stationary phase type, stationary film thickness, column inner diameter and column length.

There are two general types of columns:

1. Packed column
2. Capillary or open tubular column

3.6.3.3.1. Packed Column

Packed columns are made of glass or metal mainly stainless steel. Porous solid support material is a finely divided small uniform spherical shaped inert material, most commonly diatomaceous earth of 100-300 μm dia thickness. This solid support is coated with liquid stationary phase. Most packed columns are of 1.5- 10 m in length and have 2-4 mm internal diameter.

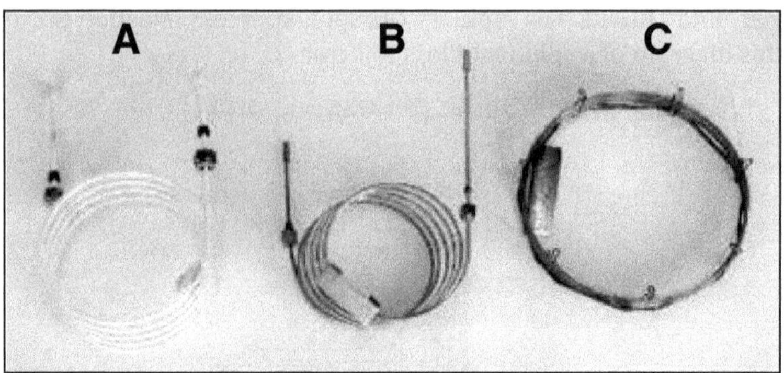

A. Packed glass column; B. Packed stainless steel column; C. Capillary fused silica column.

3.6.3.3.2. Capillary Column

Capillary columns have an internal diameter of a few tenths of a millimeter. They can be one of two types; wall-coated open tubular (WCOT) or support-coated open tubular (SCOT). Wall-coated columns consist of a capillary tube whose walls are coated with liquid stationary phase. In support-coated columns, the inner wall of the capillary is lined with a thin layer of support material such as diatomaceous earth, onto which the stationary phase has been adsorbed. SCOT columns are generally less efficient than WCOT columns. Both types of capillary column are more efficient than packed columns.

Capillary columns are made of metal, plastic, glass or fused silica. They have very small internal diameter of a few tenths of a millimeter. Capillary columns are in general more efficient than packed columns. They are two types of capillary columns:

- ☆ Wall-coated open tubular **(WCOT):** WCOT columns consist of a capillary tube whose walls are coated with liquid stationary phase.
- ☆ Support-coated open tubular **(SCOT):** SCOT columns has a capillary tube whose inner wall is lined with a thin layer of support material onto which liquid stationary phase is adsorbed. SCOT columns are generally less efficient than WCOT columns.

A more recent introduction is the Fused Silica Open Tubular (FSOT) column. They have much thinner walls than glass capillary columns. They have a polyimide coating that gives strength to the column. They are flexible and wound into coils. They have good physical strength, flexibility and low reactivity.

Film thickness of the stationary phase directly influences the retentive character and capacity of a column. The increase in film thickness leads to an increase in solute retention. The capacity of the column is directly related to film thickness, diameter and stationary phase polarity.

In 1979, a new type of WCOT column was devised - the Fused Silica Open Tubular (FSOT) column.

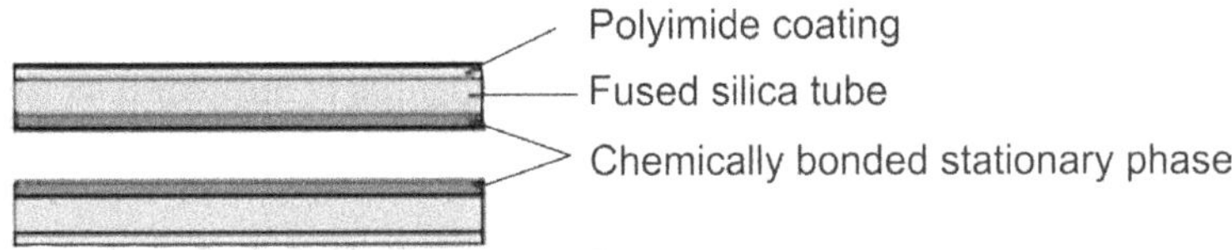

These have much thinner walls than the glass capillary columns, and are given strength by the polyimide coating. These columns are flexible and can be wound into coils. They have the advantages of physical strength, flexibility and low reactivity.

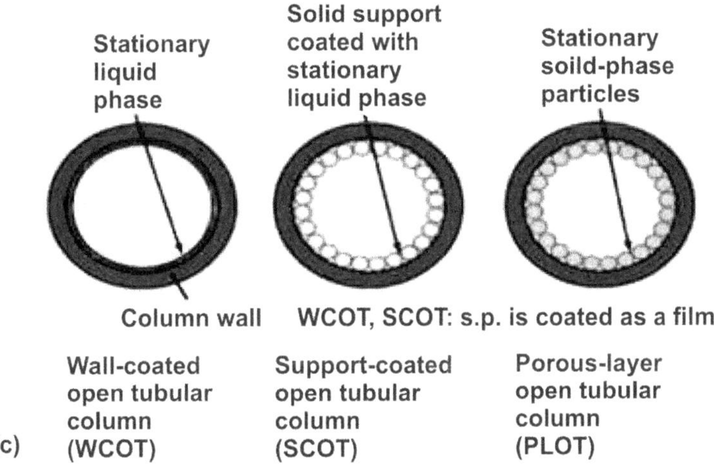

3.6.3.4. Stationery Phase

A non-polar column is better for separation of non-polar compounds and a polar column is better for the separation of polar compounds. The polarity of the stationary phase is determined by the structure of the polymer, which constitutes the stationary phase.

The two important stationery phases are:

1. Polysiloxanes
2. Polyethylene glycol

These stationery phase materials melt at high column temperature and become liquid films on the porous support materials. Stationery phase of the non-polar column is **polydimethylene-siloxane** and that of the polar column is **polyethylene glycol**.

3.6.3.4.1. Polysiloxanes

Polysiloxanes are the most common non-polar stationary phase. They are most stable, robust, and versatile. They separate the compounds according to their **boiling points** and their **different vapor pressures**. The most basic polysiloxane are 100 per cent methyl substituted. The polarity of the stationary phase increases with the substitution of methylene groups in the stationary phase such as phenyl or cyanopropyl groups. The presence of other groups is indicated as the per cent of the total number of groups *e.g.* OV-5, OV-1, *etc.*

3.6.3.4.2. Polyethylene Glycols

Polyethylene glycols are the widely used polar stationary phase with "wax". They are not substituted with any groups. They are less stable, less robust and have lower temperature limits than most polysiloxanes. In this column, the separation is due to the **polarity** and the **boiling points** of the compounds. They have shorter lifetime and more susceptible to damage by over heating or exposure to oxygen.

3.6.3.4.3. Cross-linked and Bonded Phases

In cross linked stationery phase, the individual polymer chains are linked via covalent bonds. In bonded stationary phases, the polymers are covalently bonded to the surface of the tubing. Both these forms impart enhanced **thermal** and **solvent stability** to the stationary phase. Polysiloxanes and polyethylene glycol are available in bonded and cross-linked forms. Very few of them are available in non-bonded version. It is better to use a column that has both bonded and cross-linked stationary phases, if available.

3.6.3.5. Thermostatic Oven

The column is contained in a thermostat-controlled oven. The separation of compounds in the column through the **partitioning behaviour** is dependent on the oven temperature. The optimum column temperature is fixed based on the **boiling point** of the compounds in the sample. The separation of compounds is achieved by two types of programming:

- ☆ Isothermal programming
- ☆ Gradient programming

In isothermal programming, the oven is set at same temperature and different components are eluted. In gradient programming, the oven temperature is gradually increased over time to elute the high–boiling point components with a wide range of boiling points. A temperature slightly above the average boiling point of the sample results in an elution time within 30 min. The use of minimal temperatures gives good resolution, but increase elution time. The temperature programming is useful, if the components in the sample have wide boiling points.

For precise work, column temperature must be controlled to within tenths of a degree. The optimum column temperature is dependent upon the boiling point of the sample. As a rule of thumb, a temperature slightly above the average boiling point of the sample results in an elution time of 2 - 30 minutes. Minimal temperatures give good resolution, but increase elution times. If a sample has a wide boiling range, then temperature programming can be useful. The column temperature is increased (either continuously or in steps) as separation proceeds.

Gas Chromatography (GC or GLC) is a commonly used analytic technique in many research and industrial laboratories for quality control as well as identification and quantitation of compounds in a mixture. GC is also a frequently used technique in many environmental and forensic laboratories because it allows for the detection of very small quantities. A broad variety of samples can be analyzed as long as the compounds are sufficiently thermally stable and volatile.

3.6.3.6. Detectors

There are many detectors which can be used in gas chromatography. Different detectors will give different types of selectivity. A non-selective detector responds to all compounds except the carrier gas, a selective detector responds to a range of compounds with a common physical or chemical property and a specific detector responds to a single chemical compound. Detectors can also be grouped into concentration dependant detectors and mass flow dependant detectors. The signal from a concentration dependant detector is related to the concentration of solute in the detector, and does not usually destroy the sample Dilution of with make-up gas will lower the detectors response. Mass flow dependant detectors usually destroy the sample, and the signal is related to the rate at which solute molecules enter the detector. The response of a mass flow dependant detector is unaffected by make-up gas. Have a look at this tabular summary of common GC detectors:

Detector	*Type*	*Support Gases*	*Selectivity*	*Detect-ability*	*Dynamic Range*
Flame ionization (FID)	Mass flow	Hydrogen and air	Most organic acds	100 pg	107
Thermal conductivity (TCD)	Concentration	Reference	Universal	1 ng	107
Electron capture (ECD)	Concentration	Make-up	Halides, nitrates, nitriles, peroxides, anhydrides, organometallics	50 fg	105
Nitrogen-phosphorus	Mass flow	Hydrogen and air	Nitrogen, phosphorus	10 pg	106
Flame photometric (FPD)	Mass flow	Hydrogen and air possibly oxygen	Sulphur, phosphorus, tin, boron, arsenic, germanium, selenium, chromium	100 pg	103

Detector	Type	Support Gases	Selectivity	Detect-ability	Dynamic Range
Photo-ionization (PID)	Concentration	Make-up	Aliphatics, aromatics, ketones, esters, aldehydes, amines, heterocyclics, organosulphurs, some organometallics	2 pg	107
Hall electrolytic conductivity	Mass flow	Hydrogen, oxygen	Halide, nitrogen, nitrosamine, sulphur		

3.6.3.6.1. Mass Spectrometry Detectors

Mass Spectrometer (MS) detectors are most powerful of all gas chromatography detectors. In a GC/MS system, the mass spectrometer scans the masses continuously throughout the separation. When the sample exits the chromatography column, it is passed through a transfer line into the inlet of the mass spectrometer. The sample is then ionized and fragmented, typically by an electron-impact ion source. During this process, the sample is bombarded by energetic electrons which ionize the molecule by causing them to lose an electron due to electrostatic repulsion. Further bombardment causes the ions to fragment. The ions are then passed into a mass analyzer where the ions are sorted according to their m/z value, or mass-to-charge ratio. Most ions are only singly charged.

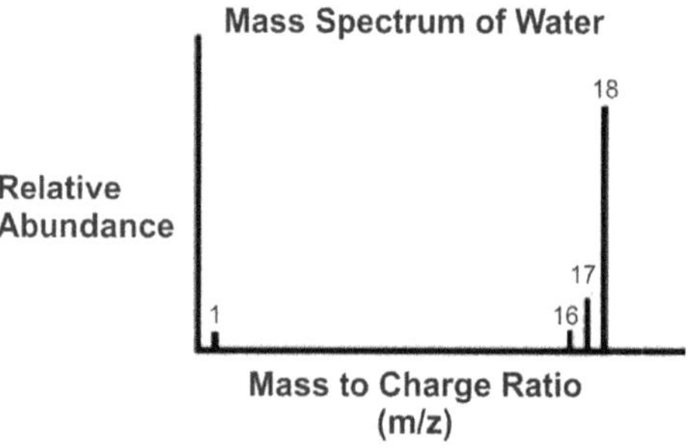

The Chromatogram will point out the retention times and the mass spectrometer will use the peaks to determine what kind of molecules are exist in the mixture. The figure below represents a typical mass spectrum of water with the absorption peaks at the appropriate m/z ratios.

One of the most common types of mass analyzer in GC/MS is the quadrupole ion-trap analyzer, which allows gaseous anions or cations to be held for long periods of time by electric and magnetic fields. A simple quadrupole ion-trap consists of a hollow ring electrode with two grounded end-cap electrodes as seen in figure. Ions are allowed into the cavity through a grid in the upper end cap. A variable radio-frequency is applied to the ring electrode and ions with an appropriate m/z value orbit around the cavity. As the radio-frequency is increased linearly, ions of a stable m/z value are ejected by mass-selective ejection in order of mass. Ions that are too heavy or too light are destabilized and their charge is neutralized upon collision with the ring electrode wall. Emitted ions then strike an electron multiplier which converts the detected ions into an electrical signal. This electrical signal is then picked up by the computer through various programs. As an end result, a chromatogram is produced representing the m/z ratio versus the abundance of the sample.

GC/MS units are advantageous because they allow for the immediate determination of the mass of the analyte and can be used to identify the components of incomplete separations. They are rugged, easy to use and can analyze the sample almost as quickly as it is eluted. The disadvantages of mass spectrometry detectors are the tendency for samples to thermally degrade before detection and the end result of obliterating the entire sample by fragmentation.

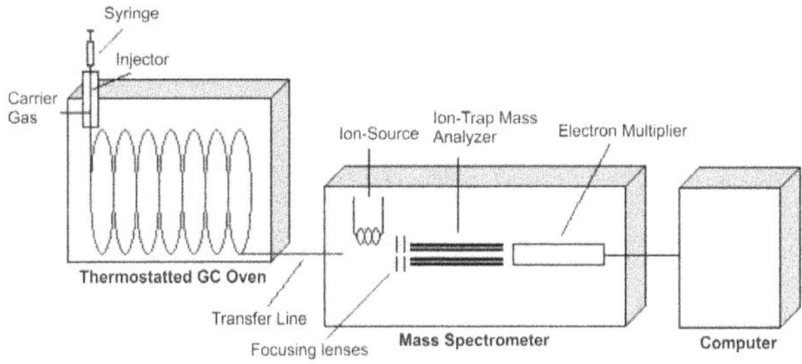

Schematic of the GC/MS System.

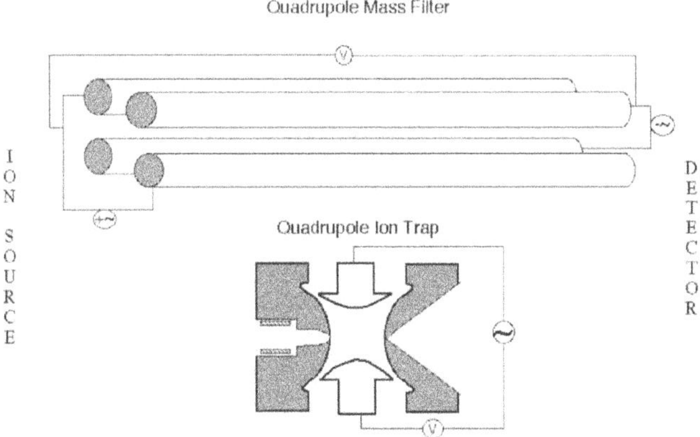

Arrangement of the Poles in Quadrupole and Ion Trap Mass Spectrometers.

3.6.3.6.2. Flame Ionization Detectors

Flame ionization detectors (FID) are the most generally applicable and most widely used detectors. In a FID, the sample is directed at an air-hydrogen flame after exiting the column. At the high temperature of the air-hydrogen flame, the sample undergoes pyrolysis, or chemical decomposition through intense heating. Pyrolized hydrocarbons release ions and electrons that carry current. A high-impedance picoammeter measures this current to monitor the sample's elution.

It is advantageous to used FID because the detector is unaffected by flow rate, non combustible gases and water. These properties allow FID high sensitivity and low noise. The unit is both reliable and relatively easy to use. However, this technique does require flammable gas and also destroys the sample.

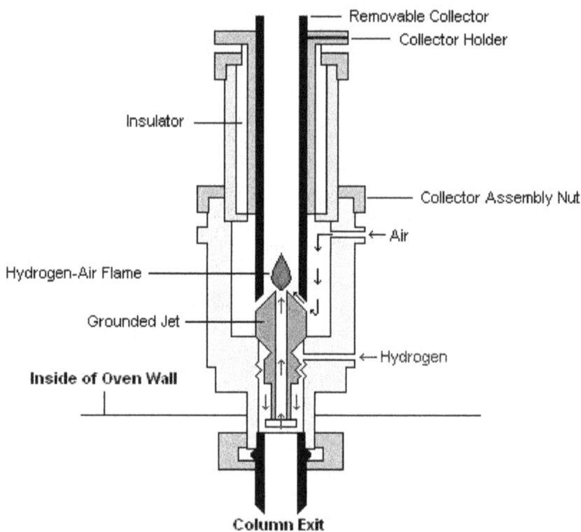

Schematic of a Typical Flame Ionization Detector.

3.6.3.6.3. Thermal Conductivity Detectors

Thermal conductivity detectors (TCD) were one the earliest detectors developed for use with gas chromatography. The TCD works by measuring the change in carrier gas thermal conductivity caused by the presence of the sample, which has a different thermal conductivity from that of the carrier gas. Their design is relatively simple, and consists of an electrically heated source that is maintained at constant power. The temperature of the source depends upon the thermal conductivities of the surrounding gases. The source is usually a thin wire made of platinum, gold or. The resistance within the wire depends upon temperature, which is dependent upon the thermal conductivity of the gas.

TCDs usually employ two detectors, one of which is used as the reference for the carrier gas and the other which monitors the thermal conductivity of the carrier gas and sample mixture. Carrier gases such as helium and hydrogen has very high thermal conductivities so the addition of even a small amount of sample is readily detected.

The advantages of TCDs are the ease and simplicity of use, the devices' broad application to inorganic and organic compounds, and the ability of the analyte to be collected after separation and detection. The greatest drawback of the TCD is the low sensitivity of the instrument in relation to other detection methods, in addition to flow rate and concentration dependency.

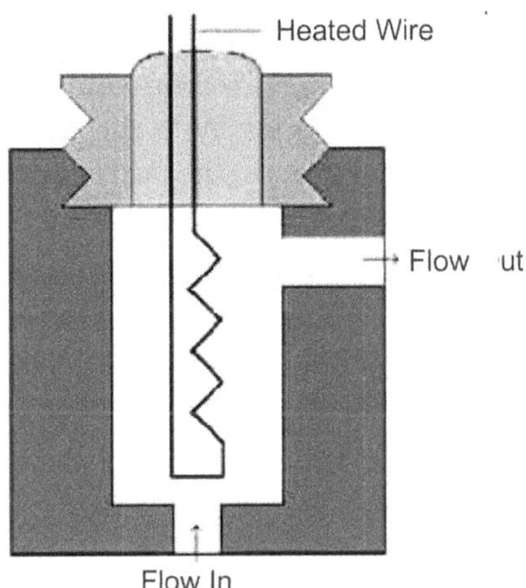

Schematic of Thermal Conductivity Detection Cell.

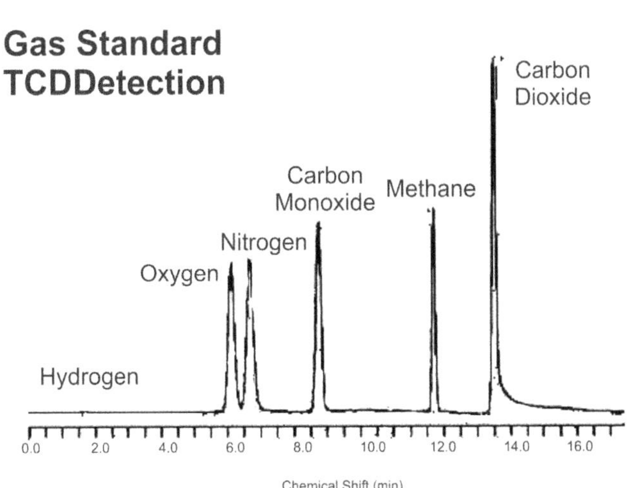

Standard Chromatogram of a Mixture of Gases.

3.6.3.6.4. Electron-capture Detectors

Electron-capture detectors (ECD) are highly selective detectors commonly used for detecting environmental samples as the device selectively detects organic compounds with moieties such as halogens, peroxides, quinones and nitro groups and gives little to no response for all other compounds. Therefore, this method is best suited in applications where traces quantities of chemicals such as pesticides are to be detected and other chromatographic methods are unfeasible.

The simplest form of ECD involves gaseous electrons from a radioactive beta emitter in an electric field. As the analyte leaves the GC column, it is passed over this beta emitter, which typically consists of nickle-63 or tritium. The electrons from the beta emitter ionize the nitrogen carrier gas and cause it to release a burst of electrons. In the absence of organic compounds, a constant standing current is maintained between two electrodes. With the addition of organic compounds with electronegative functional groups, the current decreases significantly as the functional groups capture the electrons.

The advantages of ECDs are the high selectivity and sensitivity towards certain organic species with electronegative functional groups. However, the detector has a limited signal range and is potentially dangerous owing to its radioactivity. In addition, the signal-to-noise ratio is limited by radioactive decay and the presence of 02 within the detector.

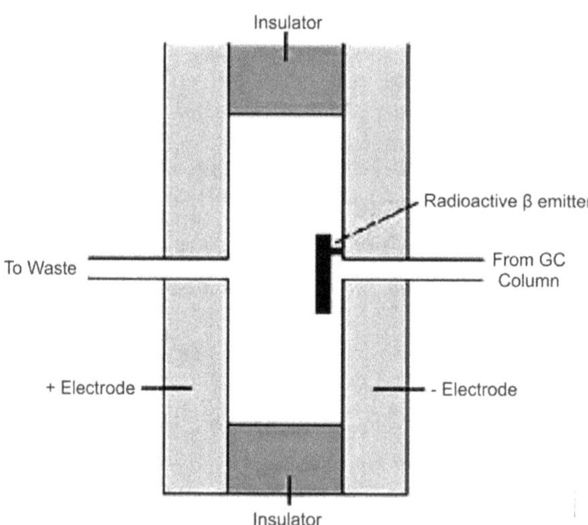

Schematic of an Electron-capture Detector.

3.6.3.6.5. Atomic Emission Detectors

Atomic emission detectors (AED), one of the newest additions to the gas chromatographer's arsenal, are element-selective detectors that utilize plasma, which is a partially ionized gas, to atomize all of the elements of a sample and excite their characteristic atomic emission spectra. AED is an extremely powerful alternative that has a wider applicability due to its based on the detection of atomic emissions. There are three ways of generating plasma: microwave-induced plasma (MIP), inductively coupled plasma (ICP) or direct current plasma (DCP). MIP is the most commonly employed form and is used with a positionable diode array to simultaneously monitor the atomic emission spectra of several elements.

The components of the Atomic emission detectors include 1) an interface for the incoming capillary GC column to induce plasma chamber, 2) a microwave chamber, 3) a cooling system, 4) a diffraction grating that associated optics, and 5) a position adjustable photodiode array interfaced to a computer.

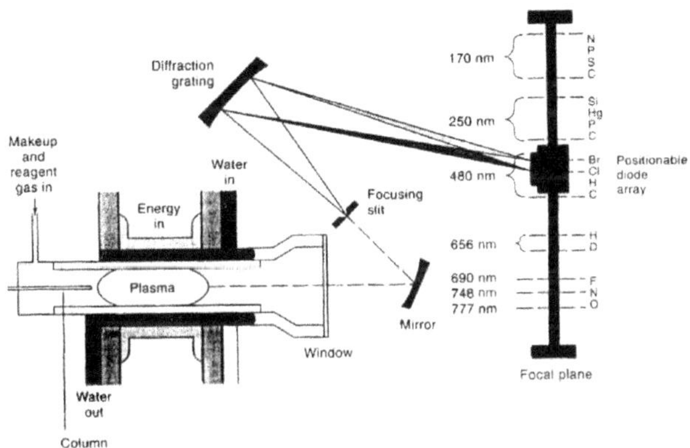

Schematic of Atomic Emission Detector.

3.6.3.6.6. GC Chemiluminescence Detectors

Chemiluminescence spectroscopy (CS) is a process in which both qualitative and quantitative properties can be be determined using the optical emission from excited chemical species. It is very similar to AES, but the difference is that it utilizes the light emitted from the energized molecules rather than just excited molecules. Moreover, chemiluminescence can occur in either the solution or gas phase whereas AES is designed for gaseous phases. The light source for chemiluminescence comes from the reactions of the chemicals such that it produces light energy as a product. This light band is used instead of a separate source of light such as a light beam.

Like other methods, CS also has its limitations and the major limitation to the detection limits of CS concerns with the use of a photomultiplier tube (PMT). A PMT requires a dark current in it to detect the light emitted from the analyte.

3.6.3.6.7. Photoionization Detectors

Another different kind of detector for GC is the photoionization detector which utilizes the properties of chemiluminescence spectroscopy. Photoionization

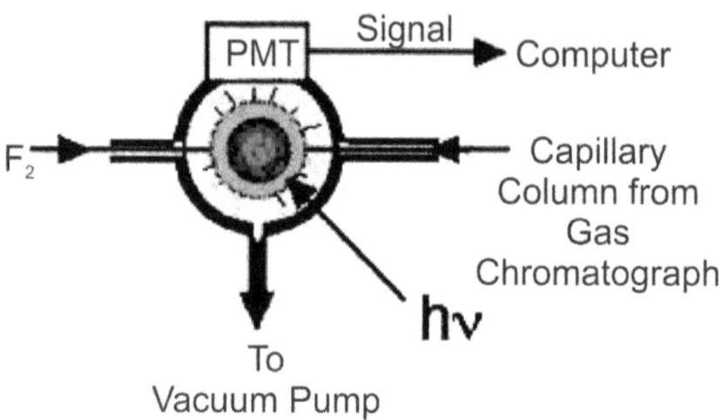

Schematic of a GC Chemiluminescence Detector.

detector (PID) is a portable vapour and gas detector that has selective determination of aromatic hydrocarbons, organo-heteroatom, inorganice species and other organic compounds. PID comprise of an ultrviolet lamp to emit photons that are absorbed by the compounds in an ionization chamber exiting from a GC column. Small fractions of the analyte molecules are actually ionized, non destructive, allowing confirmation analytical results through other detectors. In addition, PIDs are available in portable hand-held models and in a number of lamp configurations. Results are almost immediate. PID is used commonly to detect VOCs in soil, sediment, air and water, which is often used to detect contaminants in ambient air and soil. The disadvantage of PID is unable to detect certain hydrocarbon that has low molecular weight, such as methane and ethane.

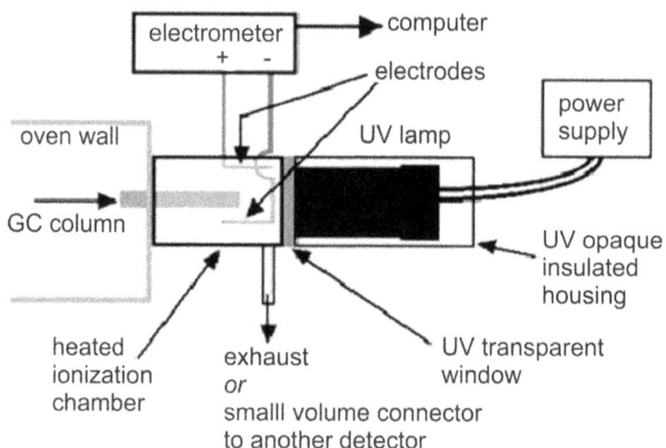

Schematic of a Photoionization Detector.

Limitations

- Not suitable for detecting semi-volatile compounds
- Only indicates if volatile organic compounds are presents.

- ✰ High concentration so methane is required for higher performance.
- ✰ Frequent calibrations are required.
- ✰ Units of parts per million ranges
- ✰ Environmental distraction, especially water vapor strong electrical field Rapid variation in temperature at the detector and naturally occurring compounds may affect instrumental signal.

3.6.3.7. Chromatography Working

Like for all other chromatographic techniques, a mobile and a stationary phase are required for this technique. The mobile phase (=carrier gas) is comprised of an inert gas *i.c.*, helium, argon, or nitrogen. The stationary phase consists of a packed column where the packing or solid support itself acts as stationary phase, or is coated with the liquid stationary phase (=high boiling polymer). Most analytical gas chromatographs use capillary columns, where the stationary phase coats the walls of a small-diameter tube directly (*i.e.*, 0.25 m film in a 0.32 mm tube).

The separation of compounds is based on the different strengths of interaction of the compounds with the stationary phase ("like-dissolves-like"-rule). The stronger the interaction is, the longer the compound interacts with the stationary phase, and the more time it takes to migrate through the column (=longer retention time). In the example above, compound X interacts stronger with the stationary phase, and therefore lacks behind compound O in its movement through the column. As a result, compound O has a much shorter retention time than compound X.

3.6.3.8. Factors Influence the Separation of Compounds

1. Boiling Point

The boiling point of a compound is often related to its polarity (see also polarity chapter). The lower the boiling point is the shorter retention time usually is because the compound will spent more time in the gas phase. That is one of the main reasons why low boiling solvents (*i.e.*, diethyl ether, dichloromethane) are used as solvents to dissolve the sample. The temperature of the column does not have to be above the boiling point because every compound has a non-zero vapor pressure at any given temperature, even solids. That is the reason why we can smell compounds like camphor (0.065 mmHg/25°C), isoborneol (0.0035 mmHg/25°C), naphthalene (0.084 mmHg/25°C), *etc.* However, their vapor pressures are fairly low compared to liquids (*i.e.*, water (24 mmHg/25°C), ethyl acetate (95 mmHg/25°C), diethyl ether (520 mmHg/25°C)).

2. The polarity of components vs. the polarity of stationary phase on column

If the polarity of the stationary phase and compound are similar, the retention time increases because the compound interacts stronger with the stationary phase. As a result, polar compounds have long retention times on polar stationary phases and shorter retention times on non-polar columns using the same temperature. Chiral stationary phases that are based on amino acid derivatives, cyclodextrins and chiral silanes are capable of separating enantiomers because one enantiomer interacts slightly stronger than the other one with the stationary phase, often due to steric effects or other very specific interactions. For instance, a cyclodextrin column is used in the determination of the enantiomeric excess in the chiral epoxidation experiment.

3. Column Temperature

A excessively high column temperature results in very short retention time but also in a very poor separation because all components mainly stay in the gas phase. However, in order for the separation to occur the components need to be able to interact with the stationary phase. If the compound does not interact with the stationary phase, the retention time will decrease. At the same time, the quality of the separation deteriorates, because the differences in retention times are not as pronounced anymore. The best separations are usually observed for temperature gradients, because the differences in polarity and in boiling points are used here

4. Carrier Gas Flow Rate

A high flow rate reduces retention times, but a poor separation would be observed as well. Like above, the components have very little time to interact with the stationary phase and are just being pushed through the column.

5. Column Length

A longer column generally improves the separation. The trade-off is that the retention time increases proportionally to the column length and a significant peak broadening will be observed as well because of increased longitudinal diffusion inside the column. One has to keep in mind that the gas molecules are not only traveling in one direction but also sideways and backwards. This broadening is inversely proportional to the flow rate. Broadening is also observed because of the finite rate of mass transfer between the phases and because the molecules are taking different paths through the column.

6. Amount of Material Injected

Ideally, the peaks in the chromatogram display a symmetric shape (Gaussian curve). If too much of the sample is injected, the peaks show a significant tailing, which causes a poorer separation. Most detectors are relatively sensitive and do not need a lot of material in order to produce a detectable signal. Strictly speaking, under standard conditions only 1-2 per cent of the compound injected into the injection port passes through the column because most GC instruments are operated in split-mode to prevent overloading of the column and the detector. The splitless mode will only be used if the sample is extremely low in concentration in terms of the analyte.

3.6.4. Applications

Gas chromatography is a physical separation method in where volatile mixtures are separated. It can be used in many different fields such as pharmaceuticals, cosmetics and even environmental toxins. Since the samples have to be volatile, human breathe, blood, saliva and other secretions containing large amounts of organic volatiles can be easily analyzed using GC. Knowing the amount of which compound is in a given sample gives a huge advantage in studying the effects of human health and of the environment as well.

Air samples can be analyzed using GC. Most of the time, air quality control units use GC coupled with FID in order to determine the components of a given air sample. Although other detectors are useful as well, FID is the most appropriate because of its sensitivity and resolution and also because it can detect very small molecules as well.

GC/MS is also another useful method which can determine the components of a given mixture using the retention times and the abundance of the samples. This method be applied to many pharmaceutical applications such as identifying the amount of chemicals in drugs. Moreover, cosmetic manufacturers also use this method to effectively measure how much of each chemical is used for their products.

3.7. High Performance Liquid Chromatography (HPLC)

High performance liquid chromatography (HPLC) is one of the major techniques widely used for analytical separation of components in a mixture using a liquid mobile phase. Martin and Synge discovered the liquid-liquid partition chromatography in 1941 and got the Nobel Prize in Chemistry in 1952. This technique has got high sensitivity, high efficiency, high resolution, and wide applicability. It is suitable for the analysis of non- volatile as well as thermally labile compounds. It is extensively used in chemical, pharmaceutical, biochemical, biomedical and environmental analyses.

3.7.1. Principle

The separation of compounds by a chromatographic column depends on the number of theoretical plates in the column, which is related to the surface area of the stationery phase. The smaller the particle size of the stationery phase, the greater is the retention of the compounds and also greater is the resistance to the flow of mobile phase. The columns with small particle size stationery phase of 5-10 μm diameter that can withstand high pressures is the basis of HPLC.

The separation of compounds is based upon stationary phase of small particle size with large surface area that are packed into short columns of 20-25 cm length or less. The mobile phase is then pumped through the system at high pressures and the separation is achieved in a few min at flow rates of 1-3 ml per min. The pumps operate at the high pressures of upto 50 MPa that are needed to force the mobile phase through the tightly packed column with a steady flow.

HPLC is a separation technique that involves the injection of a small volume of liquid sample into a tube packed with tiny particles (3 to 5 micron (μm) in diameter called the stationary phase) where individual components of the sample are moved down the packed tube (column) with a liquid (mobile phase) forced through the column by high pressure delivered by a pump. In principle, LC and HPLC work the same way except the speed, efficiency, sensitivity and ease of operation of HPLC is vastly superior. These components are separated from one another by the column packing that involves various chemical and/or physical interactions between their molecules and the packing particles. These separated components are detected at the exit of this tube (column) by a flow-through device (detector) that measures their amount. An output from this detector is called a "liquid chromatogram"

3.7.2 Instrumentation

HPLC instrument has five major parts.

- ☆ Mobile phases and its treatment system
- ☆ Pumping system
- ☆ Sample injection system
- ☆ Column
- ☆ Detector

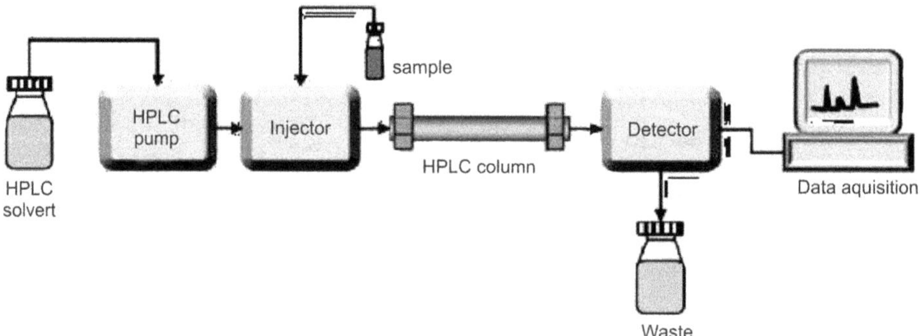

3.7.2.1. Mobile Phases

Mobile phase refers to the solvent that is continuously applied to the column or stationary phase. Pure solvents should be used as mobile phases because traces of impurities with absorbance <200 nm present in the mobile phase will affect the column and interfere with the detection system. Dust particles in the mobile phase should be removed as it interferes with detection clogs the column and damages the pumping system. A micro filter of 1-5 μm size is used before the pump to remove the dust.

3.7.2.2. Pumping System

The role of the pump is to force a liquid (called the mobile phase) through the liquid chromatograph at a specific flow rate, expressed in milliliters per min (ml/

min). Normal flow rates in HPLC are in the 1- to 2-mL/min range. Typical pumps can reach pressures in the range of 6000-9000 psi (400- to 600-bar). During the chromatographic experiment, a pump can deliver a constant mobile phase composition (isocratic) or an increasing mobile phase composition (gradient). High pressure should be applied to the mobile phase to force the solvent through the column. This is done by a motor-driven pump which can generate a pressure of upto 6000 psi or 50 MPa. Pumps should deliver pulse-free output and resistant to corrosion by a variety of solvents. They should operate at flow rates of 0.1 to 10 ml/min. Several types of pumps are available and the important ones are reciprocating piston pumps and syringe type displacement pumps.

3.7.2.2.1. Types of Pumps

a. Reciprocating Piston Pumps

They consist of a small, motor driven piston, which moves rapidly back and forth in a hydraulic chamber. These pumps can push a solvent volume between 35-400 ml. On the backstroke, the column valve is closed and the piston pulls in the solvent from the mobile phase reservoir. On the forward stroke, the pump pushes solvent out to the column from the reservoir. Advantages of this pump are small internal volume, high output pressure of up to 10,000 psi, ready adaptability to gradient elution and constant flow rates. These pumps sometimes create small pulses, which can be corrected by pulse damper.

b. Screw-driven Syringe Pumps

Screw driven syringe displacement pumps are most suitable for small-bore columns because they deliver only a finite volume of solvent. These pumps can push a solvent volume between 250-500ml. They are pulse free and ideal for isocratic elution.

3.7.2.3. Sample Injection System

The injector serves to introduce the liquid sample into the flow stream of the mobile phase. Typical sample volumes are 5- to 200-microliters (μL). The injector must also be able to withstand the high pressures of the liquid system. An auto sampler is the automatic version for when the user has many samples to analyze or when manual injection is not practical. Sample is drawn with a syringe and injected into the injector port. Injection port consists of an injection valve and a sample loop. The loop is made of metal and can hold a fixed small volume of sample (1 μl - 100 μl). Rotation of the valve opens the loop and injects the sample into the stream of the mobile phase. An auto sampler in an injection device that will inject the sample automatically in volumes of 1 μL to 1 mL.

3.7.2.4. Column

Considered the "heart of the chromatograph" the column's stationary phase separates the sample components of interest using various physical and chemical parameters. The small particles inside the column are what cause the high backpressure at normal flow rates. The pump must push hard to move the mobile phase through the column and this resistance causes a high pressure within the chromatograph.

a. Conventional Columns

Conventional columns are made of stainless steel. They are manufactured to withstand pressures up to 50MPa. They are generally 3-25 cm long and approximately 4-6 mm internal diameter to give a typical flow rate of 1-3 ml/min. At the ends of the columns, porous plug of stainless steel or Teflon are used to retain the column materials. The stationary phase is packed into a column with particles of 5-10 μm size often containing 40,000-60,000 theoretical plates/m.

b. Microbore or Open Tubular Columns

Microbore or open tubular columns have an internal diameter of 1-2 mm and length of 25-50 cm. This column is packed with 3 or 5 μm particles containing 1, 00,000 theoretical plates/m. They can sustain flow rates of 5-20 ml/min. The advantages are reduced mobile phase consumption, increased sensitivity and ideal for interfacing with mass spectrometer.

3.7.2.4.1. Stationery Phase

Stationery phase materials are spherical particles of uniform size. They are made of chemically modified silica, alumina or styrene/divinylbenzene copolymers. They exist in three forms:

- ☆ **Microporous supports:** Particle size of this support is 5-10 μm. They are most commonly used in HPLC.
- ☆ **Pellicular support:** Particles are spherical, non-porous beads with a particle size of 5 μm. They are good for the separation of proteins and large biomolecules.
- ☆ **Bonded phases:** Certain functional groups are chemically bonded onto the stationery phase such as silica. There are two types of bonded phase columns - normal and reverse phases.

1. Normal-phase Partition Chromatography

Normal phase has a highly polar stationary phase made of **triethylene glycol** or **water**. Mobile phase should be non-polar solvent such as **hexane** or **i-propyl ether**. As the compound polarity decreases or solvent polarity increases, the retention time will decrease.

2. Reversed-phase Partition Chromatography

Reverse phase is very common and has a non-polar **stationary phase** made of **hydrocarbon**. Mobile phase should be **polar** solvent such as water, ethanol, acetonitrile or tetrahydrofuran. As the compound polarity decreases or solvent polarity increases, the retention time will increase.

3.7.2.4.2. Guard Column

Guard columns are placed in front of the separating column. They serve as a protective factor that prolongs the life and usefulness of the separating column. They must be changed on a regular basis in order to optimize their protective function. They are designed to filter or remove:

- ✰ Particles that clog the separation column.
- ✰ Compounds and ions that cause "baseline drift", decrease resolution, decrease sensitivity and create false peaks.
- ✰ Compounds that cause precipitation upon contact with the stationary or mobile phase
- ✰ Compounds that co-elute, cause extraneous peaks and interfere with detection and/or quantification

3.7.2.5. Detectors

A detector can be compared to a gate watchman who verifies the visitors before permitting them entry inside a building. The chromatographic detector is capable of establishing both the identity and concentration of eluting components in the mobile phase stream. A broad range of detectors is available to meet different sample requirements. Specific detectors respond to a particular compound only and the response is independent of mobile phase composition. On the other hand the response of bulk property detectors is dependent on collective changes in composition of sample and mobile phase.

The desirable features of a detector are:

- ✰ Sensitivity towards solute over mobile phase
- ✰ Low cell volumes to minimize memory effects
- ✰ Low detector noise
- ✰ Low detection limits
- ✰ Large linear dynamic range

The detector can see (detect) the individual molecules that come out (elute) from the column. A detector serves to measure the amount of those molecules so that the chemist can quantitatively analyze the sample components. The detector provides an output to a recorder or computer that result in the liquid chromatogram (*i.e.*, the graph of the detector response). Detector detects the sample components and subsequently signals a peak on the chromatogram. There are many types of detectors.

3.7.2.5.1. Refractive Index Detector

The response is dependent on changes in refractive index of eluting compounds in the mobile phase. The mobile phase itself should have refractive index different from the sample. Gradient programming is not possible due to resulting changes in refractive index of mobile phase. The detector is less sensitive than UV-VIS detector. Temperature control is necessary as it has high temperature sensitivity. Typical applications are in Size Exclusion Chromatography. They measure the ability of sample molecules to bend or refract light, *i.e.* based on the refractive index of a particular compound. They respond to all the compounds and change the refractive index either positive or negative. They have modest sensitivity (10^{-7} gcm^{-3}) and commonly used in the analysis of carbohydrates.

3.7.2.5.2. Ultraviolet (UV) Detectors

UV-VIS Detector is the most commonly used detector. Its response is specific to a particular compound or class of compounds depending on the presence of light absorbing functional groups of eluting molecules. Some compounds which do not have such light absorbing groups can give suitable response after post column derivatization to introduce light absorbing entities. They are based upon ultraviolet-visible spectrophotometry. They are capable of measuring absorbance down to 190 nm. They use continuous flow cells with the optical path length of 10 mm, to continuously monitor the components. They have high sensitivity of 5×10^{-10} gcm^{-3}.

3.7.2.5.3. Diode Array Detectors

Incorporation of large number of diodes which serve as detector elements makes possible simultaneous monitoring of more than one absorbing component at different wavelengths. This provides benefit of time saving and cost reduction on expensive solvents. They use diode array techniques that can scan the complete spectrum of compound within seconds and display as a 3D plot on a screen in real time.

3.7.2.5.4. Fluorescent Detectors

Fluorescence detection offers greater sensitivity than a UV-VIS detector. However, the number of naturally fluorescent compounds is smaller in comparison to light absorbing compounds. This limitation is overcome by post column derivatization.

They have very high sensitivity (10-12 gcm^{-3}) than UV detector but reduced linearity. They have limited applications as relatively few compounds fluoresce. Pre-derivatization of the sample can broaden their applications.

3.7.2.5.5. Radiochemical Detectors

They involve the use of radio labeled material usually tritium (3H) or carbon-14 (14C). They detect the fluorescence associated with beta-particle ionization. They are most popular in metabolic research and has got high sensitivity limit of 10^{-9} to 10^{-10} gcm^{-3}.

3.7.2.5.6. Electrochemical Detectors

They measure compounds that undergo oxidation or reduction reactions accomplished by measuring gain or loss of electrons from samples as they pass between electrodes at a given difference in electrical potential. It has got a high sensitivity of 10^{-12} to 10^{-13} gcm^{-3}. Based on electrochemical oxidation or reduction of sample on electrode surface. It is, however, sensitive to changes in composition or flow rate of mobile phase.

3.7.2.5.7. Mass Spectroscopy Detectors

Mass spectroscopy offers very high sensitivity and selectivity. Detection is based on fragmentation of molecules by electric fields and separation on basis of mass to charge ratios of fragmented molecules. LC –MS technique has opened up new application areas due to advantages of resolution and sensitivity. The molecule is ionized and passed through a mass analyzer, where the ion current is detected. It has got detection limit of 10^{-8} to 10^{-10} $g\ cm^{-3}$.

3.7.2.5.8. Nuclear Magnetic Resonance Detectors

They give the structural information of the compounds.

3.7.2.5.9. Light-scattering Detectors

They rely on the vapourization of the compounds, evaporation of the mobile solvent and then quantification of the compounds by light scattering detected by a photodiode. The intensity of the scattered light gives the quantity of the compound present and its particle size. Light scattering detectors are useful for detection of high molecular weight molecules. After removal of mobile phase by passing through a heated zone the solute molecules are detected by light scattering depending on molecular sizes.

3.7.3. Applications

HPLC techniques can be used for the following analysis.

1. Pesticide, herbicide, fungicide and poly aromatic hydrocarbons
2. Pharmaceutical products such as antibiotics, sedatives, steroids and analgesics
3. Amino acid composition, proteins, carbohydrates, lipids, vitamins and minerals
4. Artificial sweeteners, anti oxidants, aflatoxins and additives
5. Marine biotoxins
6. Condensed aromatics, surfactants, propellants and dyes
7. Biological fluids such as bile acids, drug metabolites, urine extracts and hormones.

Chapter 4

Immunological Techniques

4.1. Radio Isotopes

4.1.1. Radioactivity

Atoms with unstable nuclei are constantly changing as a result of the imbalance of energy within the nucleus. When the nucleus loses a neutron, it gives off energy and is said to be radioactive. Radioactivity is the release of energy and matter that results from changes in the nucleus of an atom. Radioisotopes are radioactive isotopes of an element. Different isotopes of the same element have the same number of protons in their atomic nuclei but differing numbers of neutrons. They can also be defined as atoms that contain an unstable combination of neutrons and protons. there are always forces trying to push the atom nucleus apart. The nucleus is held together by something called the **binding energy**. In most cases, elements like to have an equal number of protons and neutrons because this makes them the most stable. Stable atoms have a binding energy that is strong enough to hold the protons and neutrons together. Even if an atom has an additional neutron or two it may remain stable. However, an additional neutron or two may upset the binding energy and cause the atom to become **unstable**. In an unstable atom, the nucleus changes by giving off a neutron to get back to a balanced state. As the unstable nucleus changes, it gives off radiation and is said to be radioactive. Radioactive isotopes are often called radioisotopes.

All elements with atomic numbers greater than 83 are radioisotopes meaning that these elements have unstable nuclei and are radioactive. Elements with atomic numbers of 83 and less, have isotopes (stable nucleus) and most have at least one radioisotope (unstable nucleus). As a radioisotope tries to stabilize, it may transform into a new element in a process called transmutation. We will talk about transmutation in more detail a little later.

The unstable nucleus is characterized by **excess energy** which is either imparted to a newly-created radiation particle within the nucleus, or to an atomic electron through internal conversion. This process of transformation into a new element is known as **transmutation.** The radioisotope during transmutation process undergoes radioactive decay and emits a **gamma rays** (s) and/or **subatomic particles** such as alpha and beta rays. These particles constitute **ionizing radiation.**

Radioimmunoassay (RIA) is a sensitive method for measuring very small amounts of a substance in the blood. Radioactive versions of a substance, or isotopes of the substance, are mixed with antibodies and inserted in a sample of the patient's blood. Radioimmunoassay (RIA) involves the separation of a protein (from a mixture) using the specificity of antibody - antigen binding and quantitation using radioactivity. Based on competition between unlabeled antigen and finite amount of corresponding labeled antigen for a limited number of antibody binding sites in a fixed amount of antiserum. At equilibrium in the presence of an antigen excess there will be both free antigen antigen bound to antibody.

4.1.2. Radioactive Decay

The Process of Natural Radioactive Decay

Certain naturally occurring radioactive isotopes are unstable: Their nucleus breaks apart, undergoing nuclear decay. Sometimes the product of that nuclear decay is unstable itself and undergoes nuclear decay, too. For example, when U-238 (one of the radioactive isotopes of uranium) initially decays, it produces Th-234, which decays to Pa-234. The decay continues until, finally, after a total of 14 steps, Pb-206 is produced. Pb-206 is stable, and the decay sequence, or series, stops. The nucleus has positively charged protons shoved together in an extremely small volume of space. All those protons are repelling each other. The forces that normally hold the nucleus together sometimes can't do the job, and so the nucleus breaks apart, undergoing nuclear decay.

All elements with 84 or more protons are unstable; they eventually undergo decay. Other isotopes with fewer protons in their nucleus are also radioactive. The radioactivity corresponds to the neutron/proton ratio in the atom: If the neutron/proton ratio is too high (there are too many neutrons or too few protons), the isotope is said to be neutron rich and is, therefore, unstable. If the neutron/proton ratio is too low (there are too few neutrons or too many protons), the isotope is unstable. The neutron/proton ratio for a certain element must fall within a certain range for the element to be stable. That's why some isotopes of an element are stable and others are radioactive.

There are three primary ways that naturally occurring radioactive isotopes decay:

1. Alpha particle emission
2. Beta particle emission
3. Gamma radiation emission

In addition, there are a couple of less common types of radioactive decay:

1. Positron emission
2. Electron capture

Alpha Emission

An alpha particle is defined as a positively charged particle of helium nuclei. An alpha particle is composed of two protons and two neutrons, so it can be represented as a Helium-4 atom. As an alpha particle breaks away from the nucleus of a radioactive atom, it has no electrons, so it has a +2 charge. Therefore, it's a positively charged particle of helium nuclei.

But electrons are basically free/easy to lose and easy to gain. So normally, an alpha particle is shown with no charge because it very rapidly picks up two electrons and becomes a neutral helium atom instead of an ion.

Large, heavy elements, such as uranium and thorium, tend to undergo alpha emission. This decay mode relieves the nucleus of two units of positive charge (two protons) and four units of mass (two protons + two neutrons). Each time an alpha particle is emitted; four units of mass are lost.

Radon-222 (Rn-222) is another alpha particle emitter, as shown in the following equation:

$$^{222}_{86}Rn \longrightarrow {}^{218}_{84}Po + {}^{4}_{2}He$$

alpha particle

Here, Radon-222 undergoes nuclear decay with the release of an alpha particle. The other remaining isotope must have a mass number of 218 (222 – 4) and an atomic number of 84 (86 – 2), which identifies the element as Polonium (Po).

Beta Emission

A beta particle is essentially an electron that's emitted from the nucleus. Iodine-131 (I-131), which is used in the detection and treatment of thyroid cancer, is a beta particle emitter:

$$^{131}_{53}I \longrightarrow {}^{131}_{54}Xe + {}^{0}_{-1}e$$

beta particle

Here, the Iodine-131 gives off a beta particle (an electron), leaving an isotope with a mass number of 131 (131 – 0) and an atomic number of 54 (53 – (-1)). An atomic number of 54 identifies the element as Xenon (Xe).

Notice that the mass number doesn't change in going from I-131 to Xe-131, but the atomic number increases by one. In the iodine nucleus, a neutron was converted (decayed) into a proton and an electron, and the electron was emitted from the nucleus as a beta particle.

Isotopes with a high neutron/proton ratio often undergo beta emission, because this decay mode allows the number of neutrons to be decreased by one and the number of protons to be increased by one, thus lowering the neutron/ proton ratio.

Gamma Emission

Alpha and beta particles have the characteristics of matter: They have definite masses, occupy space, and so on. However, because there is no mass change associated with gamma emission, you could refer to gamma emission as gamma radiation emission.

Gamma radiation is similar to x-rays — high energy, short wavelength radiation. Gamma radiation commonly accompanies both alpha and beta emission, but it's usually not shown in a balanced nuclear reaction.

Some isotopes, such as Cobalt-60 (Co-60), give off large amounts of gamma radiation. Co-60 is used in the radiation treatment of cancer. The medical personnel focus gamma rays on the tumor, thus destroying it.

Positron Emission

Although positron emission doesn't occur with naturally occurring radioactive isotopes, it does occur naturally in a few man-made ones. A positron is essentially an electron that has a positive charge instead of a negative charge. A positron is formed when a proton in the nucleus decays into a neutron and a positively charged electron. The position is then emitted from the nucleus. This process occurs in a few isotopes, such as

Potassium-40 (K-40), as shown in the following equation:

$$^{40}_{19}K \longrightarrow {}^{40}_{18}Ar + {}^{0}_{+1}e$$

positron

The K-40 emits the positron, leaving an element with a mass number of 40 (40 – 0) and an atomic number of 18 (19 – 1). An isotope of argon (Ar), Ar-40, has been formed.

Electron Capture

Electron capture is a rare type of nuclear decay in which an electron from the innermost energy level is captured by the nucleus. This electron combines with a proton to form a neutron. The atomic number decreases by one, but the mass number stays the same.

The following equation shows the electron capture of Polonium-204 (Po-204):

$$^{204}_{84}Po + ^{0}_{-1}e \longrightarrow ^{204}_{83}Bi + x\text{-}rays$$

The electron combines with a proton in the polonium nucleus, creating an isotope of bismuth (Bi-204). The capture of the 1s electron leaves a vacancy in the 1s orbitals. Electrons drop down to fill the vacancy, releasing energy in the X-ray portion of the electromagnetic spectrum.

4.1.3. Types of Radioisotopes

Naturally occurring radioisotopes:

1. Primordial Radioisotopes

Primordial radioisotopes originate mainly from the interiors of stars. *e.g.* Uranium and Thorium. They are still present as their half-lives are so long that they are not yet completely decayed.

2. Secondary Radioisotopes

Secondary radioisotopes are radiogenic isotopes derived from the decay of primordial radioisotopes. They have shorter half-lives than primordial radioisotopes.

3. Cosmogenic Radioisotopes

Cosmogenic isotopes are continually being formed in the atmosphere due to cosmic rays. *e.g.* Carbon-14

Artificially Produced Radioisotopes

1. Nuclear Reactors

The high flux of neutrons activates the elements placed within the nuclear reactor to produce radioisotopes. *e.g.* Thallium-201 and Iridium-192.

2. Particle Accelerators

Cyclotrons accelerate protons to bombard a target and produce positron that emits radioisotopes. *e.g.* Fluorine-18.

3. Radionuclide Generators

Radioisotopes generators contain a parent isotope produced in a nuclear reactor, that decays to produce a radioisotope. *e.g.* Technetium-99 produced Molybdenum -99.

4. Nuclear Explosions

Radioisotopes produced as an unavoidable side effect of nuclear and thermonuclear explosions.

4.1.4. Applications

Radioisotopes are applied in many ways because of their chemical properties and as a source of radiation.

- ☆ They are used as **tracers** to provide diagnostic information about a person's internal anatomy and function of specific organs in **tomography.**
- ☆ They are used to treat **hemopoietic** forms of tumors.
- ☆ They are used to sterilize **syringes** and other medical equipments eg.gamma rays.
- ☆ They are used to **stop sprouting** of root crops after harvesting, to **kill parasites** and pests, and to **control ripening** of stored fruit and vegetables.
- ☆ They are used to trace chemical and physiological processes occurring in living organisms, such as DNA replication or amino acid transport.
- ☆ They help to examine welds, to detect leaks, erosion and corrosion of metals.
- ☆ They are used to trace and analyze **pollutants**, to study the movement of surface water, and to measure water runoffs from rain and snow, as well as the flow rates of streams and rivers.
- ☆ They are used to measure ages of rocks, minerals, and fossil materials.

4.2. Radioimmunoassay

The radioimmunoassay technique, as the name implies, achieves sensitivity through the use of radionuclides and specificity that is uniquely associated with immune chemical reactions. Yalow and Berson were awarded the Nobel prize for pioneering radioimmunoassay when they evolved radioimmunological assay for insulin.

Principle of Radioimmunoassay

The radioimmunoassay technique is based on the isotope dilution principle, along with the use of a specific antibody to bind to a portion of the substance to be measured. If an antigen (for example, a hormone) is mixed with a specific antibody to that substance, an interaction will occur, forming an antigen/antibody complex that is chemically different from either the antigen or the antibody. If there is insufficient antibody to complex all the antigen present, mixing of the antibody with a known amount of isotopically labeled antigen along with an unknown amount of labeled antigen allows quantitation of the unlabeled antigen.

Radioimmunoassay (RIA) is a highly sensitive and specific assay used to determine the concentration of an antibody or an antigen by exploiting the competition between radio labeled and unlabeled substances in an antigen-antibody reaction. This assay was developed by Rosalyn Yalow and Solomon Aaron Berson in 1960 to detect the concentration of insulin in plasma. RIA technique requires specialized equipments and special precautions as radioactive substances are employed.

4.2.1. Technique

- ✰ A known quantity of an antigen is made radioactive by labelling it with gamma-radioactive isotopes.
- ✰ The radio labeled antigen is then mixed with a known quantity of specific antibody and they will chemically bind to one another.
- ✰ Then, the sample containing an unknown quantity of the same antigen or unlabeled antigen is then added.
- ✰ The unlabeled antigen also competes with the radio labeled antigen for the antibody binding sites.
- ✰ If the concentration of unlabeled antigen is increased, it displaces the radio labeled antigen and binds more to the antibody
- ✰ The ratio of antibody-bound radio labeled antigen to free radio labeled antigen is then reduced.
- ✰ The bound antigens are then separated from the unbound antigens.
- ✰ The radioactivity of the free antigen remaining in the supernatant is then measured.
- ✰ A binding curve is then generated with the known standards and the amount of antigen can be derived.

4.2.2. Separation of Bound Antigen from Free Antigen

There are several ways for the separation of bound antigen from free antigen.

- ✰ The antigen-antibody complexes are precipitated by the addition of a "second" antibody directed against the first antibody. *e.g.* If a rabbit IgG is used to bind the antigen, the complex can be precipitated by the addition of an antirabbit-IgG antiserum
- ✰ The antigen-specific antibodies can be coupled to the inner walls of a test tube. After incubation, the contents containing "free" antigen are removed and then the tube with bound antigen is washed and the radioactivity of both can be measured.
- ✰ The antigen-specific antibodies can be coupled to particles, like Sephadex. The bound antigen can be separated as pellets from the free antigen in the supernatant fluid by centrifugation of the reaction mixture.

4.2.3. Applications

The technique of radioimmunoassay is used in many areas such as blood banking, diagnosis of allergies and endocrinology. Some of the specific applications are

- Blood banking
- Diagnosis of allergies
- Endocrinology
- Assay of hormone levels in plasma.
- Presence of hepatitis B surface antigen (HBsAg) in blood.
- Presence of bacterial toxins in food.

4.3. ELISA

4.3.1. Introduction

The ELISA, Enzyme linked Immunosorbent assay, also sometimes known as EIA *i.e.* Enzyme Immuno Assay is a rapid test used for detecting and quantifying antibodies or antigens in specimen against viruses, bacteria and other materials. This method is used to detect/diagnose infectious, auto immune and other diseases. ELISA is carried out on solid matrix, in 96 well microtitre plates or strips (12 wells or 8 wells each) made of polystyrene/commercially available coombs/cartridges (Rapid ELISA). The protein antigen is affixed to any of above mentioned surfaces, and then a specific antibody is applied over the surface so that it can bind to the antigen. This antibody is linked to an enzyme, and in the final step a substancc/ substrate specific for the enzyme is added that the enzyme can convert to some detectable signal, most commonly a colour change. Performing ELISA, like other antigen antibody reactions, involves at least one antibody with specificity for a particular antigen. Depending upon whether we want to detect antibody or antigen the type of ELISA will vary. For example the antigen in a sample can be detected either by direct ELISA, sandwich ELISA or competitive ELISA and antibody is usually detected by indirect ELISA. Another thing to note is that between each step of assay, whatever the ELISA format is, the plate is washed with a mild detergent solution to remove any un reacted proteins or antibodies. After the final wash step, the plate is developed by adding an enzymatic substrate to produce a visible signal, which indicates the quantity of antigen/antibody in the sample and the reaction is read either with naked eye or with an ELISA Reader. The result is expressed as OD (Optic Density) value (Reader) or titre.

The only option for conducting an immunoassay was Radioimmunoassay, a technique which used radioactively labeled antigens or antibodies for diagnostics before the advent of ELISA. Rosalyn Sussman Yalow and Solomon Berson published a paper in 1960 on Radioimmunoassay. Radioactivity was the reporter providing the signal which indicated the presence of analyte being sought in the sample. However, radioactivity posed a potential health threat, so safer reporter alternatives were sought. One such alternative reporter tested was enzyme peroxidase which reacted with appropriate substrates (such as ABTS or 3, 3', 5, 5'-Tetramethylbenzidine) to produce a change in colour, which could be used as a signal. The next step in history was the linking of an antibody/antigen to the enzyme. This linking process was independently developed by two different scientists Stratis Avrameas and G.B. Pierce. The next development was the idea of fixing the antigen/antibody to prepare the immunosorbent surface as it is necessary to remove any unbound antibody or antigen by washing. Wide and Porath developed the technique and published the same in 1966. Perlmann, Schuurs, Engvall, and van Weemen were honored with the German scientific award of the "Biochemische Analytik" for their inventions in 1976. EIA/ELISA used the principles of conventional radioimmunoassay, with the key difference that the antibodies are labeled with an enzyme, rather than radioisotopes. Now other reporter molecules like fluorogenic, electro chemi luminescent, and real-time PCR and appropriate signal systems have been developed and available in Medicine for diagnostics. Well established in vitro diagnostic industries has developed based on these technologies and are marketing a huge number of EIAs/ ELISAs.

4.3.2. Technique

An antigen is immobilized on a solid support usually a polystyrene microtiter plate. Immobilization is done either non-specifically, through adsorption to the surface or specifically, through capture by another antibody specific to the same antigen. The detection antibody is then added to form a complex with the antigen. The detection antibody can covalently link to an enzyme, or can itself be detected by a secondary antibody which is linked to an enzyme through bioconjugation. The plate is typically washed with a mild detergent solution between each step to remove any antigen or antibodies that are not specifically bound. The plate is then developed by adding an enzymatic substrate to produce a visible signal, which indicates the quantity of antigen in the sample.

4.3.3. Components of ELISA

1. Antigen

Any substance that stimulates an immune response is known as antigen. It is the target protein in the sample that binds to the antibody.

2. Antibody

An antibody is a protein made in response to an antigen. Each antibody binds in the epitope region of its antigen. Two types of antibodies *viz.* monoclonal or polyclonal can be produced and used as the capture and detection antibodies. The monoclonal antibodies are used as detection antibody as they have mono specificity toward a single epitope and allows fine detection and quantitation of small differences in antigen.

3. Enzyme Conjugate

An enzyme conjugate is an antibody joined with an enzyme. The joining of the enzyme to antibody is called **conjugation**. The detection enzyme can be linked directly to the primary antibody or introduced through a secondary antibody that recognizes the primary antibody. The most commonly used enzyme labels are horseradish peroxidase (HRP) and alkaline phosphatase (AP).

4. Substrate

A substrate is a compound that undergoes change. The substrate binds to active sites on the surface of enzyme and gets converted by the change of color. The choice of substrate depends upon the assay sensitivity and instrumentation available for signal-detection such as spectrophotometer, fluorometer or luminometer. The common ELISA substrates are:

- ☆ **PNPP** (p-Nitrophenyl phosphate, Disodium Salt) is used to detect alkaline phosphatase. It produces a yellow water-soluble reaction product that absorbs light at 405 nm.
- ☆ **ABTS** (2,2'-Azinobis [3-ethylbenzothiazoline-6-sulfonic acid]-diammonium salt) is used to detect horse radish peroxidase (HRP). It produces a water-soluble green end reaction product which has two major absorbance peaks, 410 nm and 650 nm. ABTS is less sensitive than OPD and TMB
- ☆ **OPD** (o-phenylenediamine dihydrochloride) is used to detect horse radish peroxidase and yields a water soluble yellow-orange reaction with an absorbance maximum of 492 nm.
- ☆ **TMB** (3,3',5,5'-tetramethylbenzidine) is used to detect horse radish peroxidase. It yields a blue color with absorbance maxima at 370 nm and 652 nm. It is very sensitive.

4.3.4. Classification of ELISA

ELISA are classified according to the way in which the target antigens are captured and detected.

Based on capture, there are two methods.

1. **Antigen capture ELISA** – The target antigen is captured by the microplate
2. **Antibody capture ELISA** – The target antigen is captured by an antibody attached to the microplate

Based on the detection, there are two methods

1. **Direct ELISA** - The enzyme conjugate binds directly to the antigen
2. **Indirect ELISA** – The enzyme conjugate binds indirectly to the antigen

Apart from these methods, there are two special ELISA methods.

1. Competitive ELISA
2. Reverse ELISA

4.3.4.1. Direct ELISA

In direct ELISA, a labeled primary antibody is allowed to react directly with the antigen. This is performed with antigen that is directly immobilized on the assay plate. This is not widely used, however, used in immuno histochemical staining of tissues and cells.

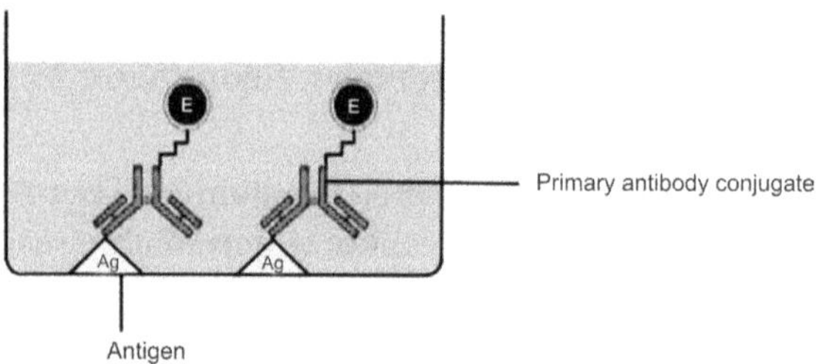

Direct Antigen Capture.

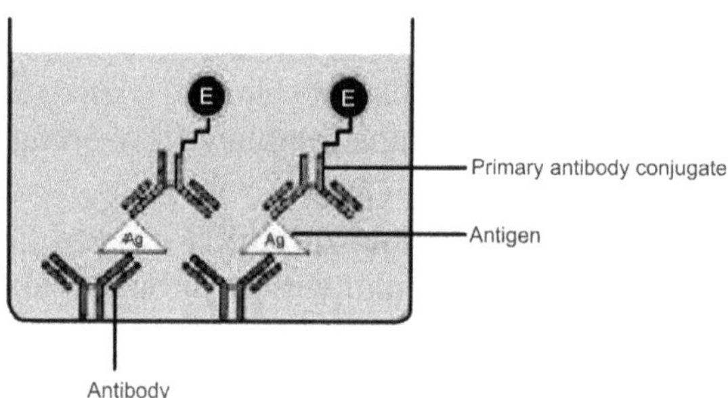

Direct Antibody Capture.

4.3.4.2. Indirect ELISA

In indirect ELISA, a labeled secondary antibody is used for the detection of the antigen, that is directly immobilized on the assay plate. The antigen first binds to a primary antibody, which then binds to the labeled secondary antibody.The secondary antibody has specificity for the primary antibody but not for the capture antibody. This assay is therefore not specific for the antigen. The indirect ELISA is the most widely used method.

Double Antibody Sandwich (DAS) ELISA

This is the most powerful direct ELISA method. In this, the capture antibodies are coated to microtiter plate wells. The antigen is then added and allowed to bind with capture antibody. The primary antibody with enzyme conjugate is then added and it binds directly to the antigen to form the double antibody sandwich. It is called a "sandwich" ELISA because the target antigen is bound between two antibodies, the capture antibody and the detection antibody. It is more sensitive and robust.

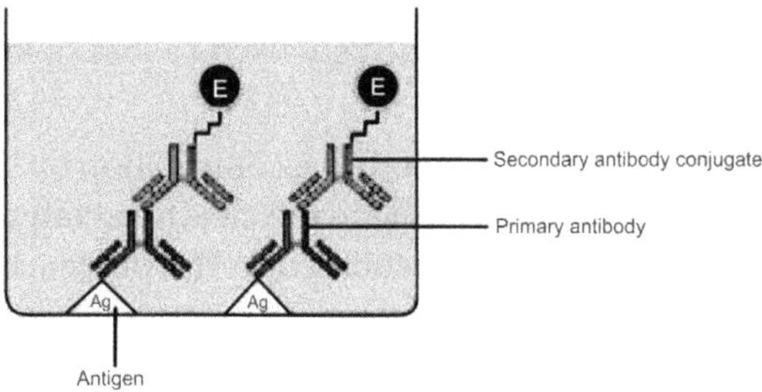

Indirect Antigen ELISA.

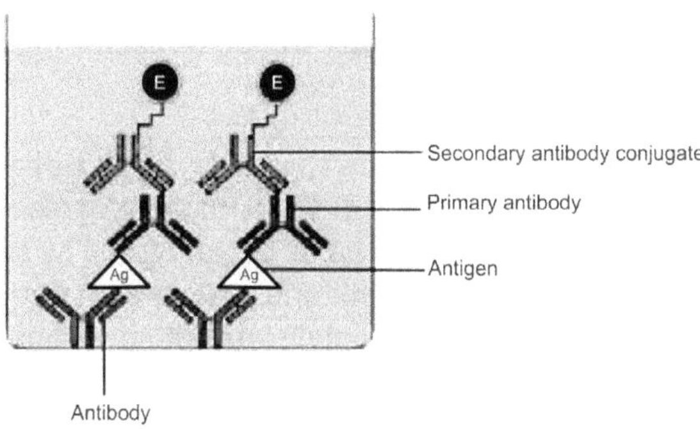

Indirect Antibody ELISA.

Indirect Triple Antibody Sandwich (TAS) ELISA

In TAS ELISA, the target antigen is captured by an antibody attached to the microtiter plate. The primary antibody is then added and allowed to bind with the target antigen. The excess antibody is washed away. The capture and primary antibodies should be obtained from different host species (*e.g.*, mouse IgG and rabbit IgG). Then, the secondary antibody with enzyme conjugate is added and allowed to bind indirectly to the antigen.

4.3.4.3. Competitive ELISA

The competitive ELISA is used to quantify antigen using competitive method. This method is used to assay small molecules usually an antigen with one epitope such as hormones in blood. There are competitive ELISA kits available with enzyme conjugated antigen and enzyme conjugated antibody.

In enzyme conjugated antigen assay, there is a competition between natural target antigen and enzyme conjugated antigen. The target antigen and the enzyme conjugated antigen are mixed in a microplate well. Both of them compete for the primary antibody sites. If the natural antigen is more, it displaces the enzyme conjugated antigen more leading to their reduction. The substrate is then added and colour development is recorded. If more natural antigen is bound then colour development is low. This method is routinely used to test thyrozine hormone in blood.

In the enzyme conjugated assay, there is a competition by the enzyme-linked secondary antibody with the sample antigen which is associated with the primary antibody. The unlabeled antibody is incubated in the presence of natural target antigen. The bound antibody/antigen complexes are then added to an antigen coated microtiter well. The plate is washed to remove the unbound antibody. The more antigens in the sample, the fewer antibodies will be able to bind to the antigen in the well. The enzyme conjugated secondary antibody, specific to the primary antibody is added. Then, a substrate is added and remaining enzymes elicit a chromogenic or fluorescent signal. For competitive ELISA, the higher the original antigen concentration, the signal will be weaker.

4.3.4.4. Reverse ELISA

Reverse ELISA is a new technique that uses a solid phase made up of an immunosorbent polystyrene rod with 4-12 protruding ogives. The entire device is immersed in a test tube containing the target antigen. The steps such washing, incubation in conjugate and chromogenic substrate are carried out by dipping the **ogives** in microwells of standard microplates pre-filled with reagents.

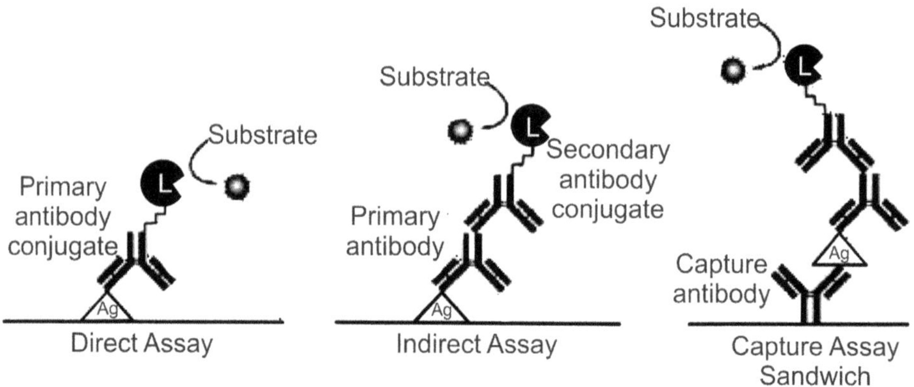

4.3.5. ELISA Plate Reader

ELISA reader is a machine that measures color density in the microtiter plate well at a specific wavelength of light. The light is cast from the bottom of the well through the sample. The wells with no color absorb very little light and the wells with color change absorb more light. The color intensity is measured as the optical density (OD).

4.3.6. Applications

The discovery of ELISA many advances have been made to make this assay more and more sensitive and specific. ELISA is employed in Medicine to detect and diagnose microbial infectious diseases, autoimmune diseases, detection of antigen in a given sample.

- ELISA can be applied to detect and measure antibody in serum against
- viruses, bacteria, parasites
- Detection of plant pathogens, mother stock and seed/leaf purity in the field of agriculture
- Determination of the level of antibodies in faecal content.
- ELISA has also been used in home pregnancy test (Rapid ELISA)
- ELISA is used in food industry to detect potential food allergens such as milk, peanuts, walnuts, almonds, and eggs.
- ELISA can also be used in toxicology as rapid presumptive screen for certain classes of drugs.
- The ELISA was widely used in various areas such as immunology,
- Biological Pharmacy, Diagnostic industry, and so on.

Chapter 5

Centrifugation

5.1. Principles of Centrifugation

5.1.1. Introduction

Centrifugation is one of the most important and widely applied research techniques in biochemistry, medicine, cellular and molecular biology for the isolation of cells and viruses, separation of subcellular organelles and isolation of macromolecules such as DNA, RNA, proteins or lipids. Centrifugation is carried out by a centrifuge that uses centrifugal force or g-force to isolate suspended particles from their surrounding medium. Biological centrifugation is a process that uses centrifugal force to separate and purify mixtures of biological particles in a liquid medium. It is a key technique for isolating and analyzing cells, sub-cellular fractions, supramolecular complexes and isolated macromolecules such as proteins or nucleic acids. The development of the first analytical ultracentrifuge by Svedberg in the late 1920s and the technical refinement of the preparative centrifugation technique by Claude and colleagues in the 1940s positioned centrifugation technology at the centre of biological and biomedical research for many decades. Today, centrifugation techniques represent a critical tool for modern biochemistry and are employed in almost all invasive subcellular studies. While analytical centrifugation is mainly concerned with the study of purified macromolecules or isolated supramolecular assemblies, preparative centrifugation methodology is devoted to the actual separation of tissues, cells, subcellular structures, membrane vesicles and other particles of biochemical interest.

Most undergraduate students will be exposed to preparative centrifugation protocols during practical classes and might also experience a demonstration of analytical centrifugation techniques. This chapter is accordingly divided into a short introduction into the theoretical background of sedimentation, an overview of practical aspects of using centrifuges in the biochemical laboratory, an outline of preparative centrifugation and a description of the usefulness of ultracentrifugation techniques in the biochemical characterization of macromolecules. To aid in the understanding of the basic principles of centrifugation, the general design of various rotors and separation processes is diagrammatically represented. Often the learning process of undergraduate students is hampered by the lack of a proper linkage between theoretical knowledge and practical applications. To overcome this problem, the description of preparative centrifugation techniques is accompanied by an explanatory flow chart and the detailed discussion of the subcellular fractionation protocol of a specific tissue preparation. Taking the isolation of fractions from skeletal muscle homogenates as an example, the rationale behind individual preparative steps is explained.

5.1.2. Basic Principles of Sedimentation

Particles or cells in a liquid suspension get sediment at the bottom of a container due to gravity. The time required for such separation is usually very long. The sedimentation effect is mainly influenced by the Earth's **gravitational field** (g=98.1 cms^{-2}).

Therefore, small particles will not separate under normal gravitational field and require **centrifugal force**. This force increases the rate of sedimentation of particles in a **centrifugal field.**

If a particle (m) in a centrifuge tube filled with a liquid is in centrifugal field, the particle (m) is acted by three forces:

- ☆ F_G: the centrifugal force
- ☆ F_B: the buoyant force
- ☆ F_D: the frictional force between particle and liquid

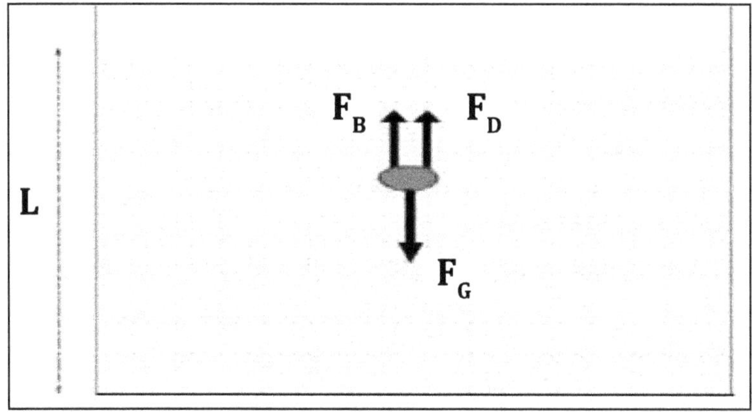

As the particles moves towards the bottom of the centrifuge tube, its velocity (v) will increase due to the increase in the radial distance (r). At the same time, the particles also encounter a frictional drag proportional to their velocity. The frictional force (F_D) of a particle is the product of its velocity and its frictional coefficient (f). The frictional force always acts in the opposite direction to sedimentation.

The rate of sedimentation is dependent upon the applied **centrifugal field** (cms^{-2}), **G**. The **centrifugal field** is determined by the radial distance (r) of the particle from the axis of rotation (in cm) and the square of angular velocity (ω) of the rotor (in radians per second).

$$G = \omega^2 r$$

The **average angular velocity (ω)** is defined as the ratio of the angular displacement in the given time interval. One radian (1 rad) represents the angle subtended at the centre of a circle by an arc with a length equal to the radius of the circle. One revolution of the rotor can be expressed as 2π radians since 360° equals 2π radians. The angular velocity in rads per seconds of the rotor can be expressed in terms of **rotor speed, s:**

$$\omega = 2\pi s/60$$

And therefore the centrifugal field can be expressed as:

$$G = 4\pi^2 (rev.min^{-1})^2 r/3600 \text{ or } 4\pi^2 s^2 r/3600$$

The centrifugal field is generally expressed in multiples of the gravitational field, g (98.1 cms^{-1}). The **relative centrifugal field, g** (RCF) is the ratio of the centrifugal acceleration at a specified radius and the speed to the standard acceleration of gravity. It can be calculated as follows:

$$RCF = 4\pi^2 (rev.min^{-1})^2 r/3600 \times 98.1 = G/g$$

RCF units are dimensionless and revolutions per minute are expressed as r.p.m:

$$RCF = 1.12 \times 10^{-5} \times r.p.m^2 \times r$$

The sedimentation rate or velocity of the biological particle is expressed as its **sedimentation coefficient (s)**:

$$s = v/\omega^2 r$$

The sedimentation coefficients of biological macromolecules are relatively small and are usually expressed as **Svedberg units, S.** One Svedberg units equals 10-13s.

5.1.3. Types of Centrifuges

Centrifugation techniques take a central position in modern biochemical, cellular and molecular biological studies. Depending on the particular application, centrifuges differ in their overall design and size. However, a common feature in all centrifuges is the central motor that spins a rotor containing the samples to be separated. The biological medium is chosen for the specific centrifugal application and may differ considerably between preparative and analytical approaches.

The most obvious differences between centrifuges are:

- The maximum speed at which biological specimens are subjected to increased sedimentation;
- The presence or absence of a vacuum;
- The potential for refrigeration or general manipulation of the temperature during a centrifugation run; and
- The maximum volume of samples and capacity for individual centrifugation tubes.

Many different types of centrifuges are commercially available including:

- Large-capacity low-speed preparative centrifuges;
- Refrigerated high-speed preparative centrifuges;
- Analytical ultracentrifuges;
- Preparative ultracentrifuges;
- Large-scale clinical centrifuges; and
- Small-scale laboratory microfuges.

Some large-volume centrifuge models are quite demanding on space and also generate considerable amounts of heat and noise, and are therefore often centrally positioned in special instrument rooms in biochemistry departments. However, the development of small-capacity bench-top centrifuges for biochemical applications, even in the case of ultracentrifuges, has led to the introduction of these models in many individual research laboratories.

Classification of Centrifuges Based on the Speed of Rotation

Centrifuge Classes	*Low Speed*	*High-Speed*	*Ultra/Micro-ultra*
Maximum speed (rpm x 10^3)	10	28	100/150
Maximum RCF (x10^3)	7	100	800/900
Pelleting applications			
Bacteria	Yes	Yes	Yes
Animal and Plant cells	Yes	Yes	Yes
Nuclei	Yes	Yes	Yes
Precipitates	Some	Most	Yes
Membrane fractions	Some	Some	Yes
Ribosomes / Polysomes	-	-	Yes
Macromolecules	-	-	Yes
Viruses	-	Most	Yes

5.1.3.1. Differential Centrifugation

In differential centrifugation, separation is achieved based on the size of the particles. This type of separation is commonly used in simple pelleting and in obtaining partially-pure preparation of subcellular organelles and macromolecules. The tissues or cells are first disrupted to release their internal contents and this crude mixture is referred as **cell homogenate**. The large particles get sediment faster than the smaller ones during centrifugation and different subcellular fractions are obtained.

When a cell homogenate is centrifuged at 600g for 10 min, unbroken cells and heavy nuclei pellet settle to the bottom of the tube. The supernatant is further centrifuged at 10,000g for 20 min to pellet subcellular organelles of intermediate velocities such as mitochondria, lysosomes, and microbodies. These partially-purified organelles can be further purified using density gradient separation.

5.1.3.2. Density Gradient Centrifugation

Density gradient centrifugation is the preferred method to purify subcellular organelles and macromolecules. Density gradients are generated by placing layer after layer of gradient media such as sucrose or caesium chloride (CsCl) in a tube with the heaviest layer at the bottom and the lightest at the top either in a discontinuous or continuous mode. Cell fraction to be separated is placed on top of the layer, centrifuged and separated according to their size or densities. The density gradient separation is classified further into two categories: rate-zonal (size) separation and isopycnic (density) separation.

5.1.3.2.1. Rate Zonal (Size) Separation

Rate-zonal separation depends on particle size and mass instead of particle density for sedimentation. Most common applications include separation of cellular organelles such as endosomes or antibodies. Antibody classes generally have very similar densities but different masses; and can be separated based on density.

5.1.3.2.2. Isopycnic (Density) Separation

Isopycnic separation depends on particle density. A particle of a particular density will sink during centrifugation until a position is reached where the density of the surrounding solution is exactly the same as the density of the particle. Once this quasi-equilibrium is achieved, the length of centrifugation does not have any influence on the migration of the particle. Most common application is separation of nucleic acids in a caesium chloride (CsCl) gradient and cell organelles in a surcose gradient. A variety of gradient media can be used for isopycnic separations listed in Table

Gradient	*Cells*	*Viruses*	*Organelles*	*Nucleo Proteins*	*Macromolecules*
Sugar (sucrose)	+	+++	+++	+	-
Polysaccharides (Ficoll)	++	++	++	-	-
Colloidal silica (Percoll)	+++	+	+++	-	-
Iodinated media (Nycodenz)	++++	++	++++	+++	+
Alkaline metal salts (CsCl)	-	++	-	++	++++

++++ excellent; +++ good; ++ good for some applications; + limited use; - unsatisfactory.

5.1.4. Rotor Categories

To illustrate the difference in design of fixed-angle rotors, vertical tube rotors and swinging-bucket rotors, Figure outlines cross-sectional diagrams of these three main types of rotors. Companies usually name rotors according to their type of design, the maximum allowable speed and sometimes the material composition. Depending on the use in a simple low-speed centrifuge, a high-speed centrifuge, or an ultracentrifuge, different centrifugal forces are encountered by a spinning rotor. Accordingly different types of rotors are made from different materials. Low-speed rotors are usually made of steel or brass, while high-speed rotors consist of aluminium, titanium or fibre-reinforced composites. The exterior of specific rotors might be finished with protective paints. For example, rotors for ultracentrifugation made out of titanium alloy are covered with a polyurethane layer. Aluminium rotors are protected from corrosion by an electrochemically formed tough layer of aluminium oxide. In order to avoid damaging these protective layers, care should be taken during rotor handling.

Rotors can be broadly classified into three common categories.

- ✰ Swinging-bucket rotors
- ✰ Fixed-angle rotors
- ✰ Vertical rotors

Low-speed rotors are usually made of steel or brass, while high-speed rotors are made of aluminum, titanium or fibre reinforced composites.

5.1.4.1. Swinging Bucket Rotor

In swinging bucket rotor, the sample tubes are loaded into individual buckets that hang vertically while the rotor is at rest. When the rotor begins to rotate, the buckets swing out to a horizontal position. This rotor is useful when samples are to be separated based on density gradients. This rotor is inefficient for pelleting.

5.1.4.2. Fixed Angle Rotor

In fixed-angle rotor, the sample tubes are held fixed at the angle of the rotor cavity. When the rotor begins to rotate, the solution in the tubes reorients to effect the separation. This rotor is most commonly used for pelleting by differential separation of biological particles. It is also useful for isopycnic separations of **macromolecules** such as nucleic acids.

5.1.4.3. Vertical Rotor

In vertical rotor, sealed tubes are held parallel to the axis of rotation. Samples are not separated down the length of the centrifuge tube, but across the diameter of the tube. The isopycnic separation time is shorter in this rotor as compared to swinging bucket rotor. This rotor is not suitable for pelleting applications but is most efficient for isopycnic separations due to the short pathlength. Most common applications are isolation of plasmid DNA, RNA, and lipoproteins.

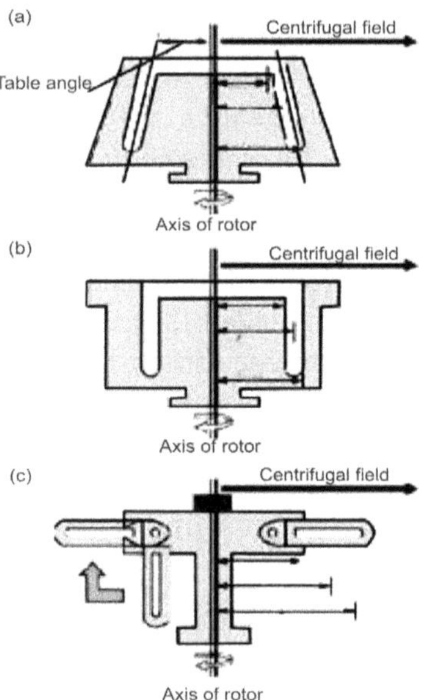

Design of the Three Main Types of Rotors Used in Routine Biochemical Centrifugation Techniques. Cross-sectional diagram of a fixed-angle rotor (a), a vertical tube rotor (b), and a swinging-bucket rotor (c). A fourth type of rotor is represented by the class of near-vertical rotors (not shown).

5.1.5. Selection of Centrifuge Tubes

Centrifuge tubes should prevent sample leakage or loss, ensure chemical compatibility and allow easy sample recovery. The major factors in the selection of a tube material include clarity, chemical resistance and sealing mechanism. The popular materials include polypropylene (PP), polyallomer (PA), polycarbonate (PC) and polyethylene terephtalate (PET). PP and PA tubes are opaque but have good chemical resistance, whereas PC and PET tubes are transparent but have poor chemical resistance. It is recommended to use thin-walled sealed tubes in a fixed angle or vertical rotor.

5.2. Ultracentrifugation

5.2.1. Ultracentrifugation

The ultracentrifuge is a centrifuge optimized for spinning a rotor at very high speeds, capable of generating acceleration as high as 2,000,000 g (approx. 19 600 km/s^2). There are two kinds of ultracentrifuges, the preparative, and the analytical ultracentrifuge. It is an important tool in biochemical research is the centrifuge, which through rapid spinning imposes high centrifugal forces on suspended

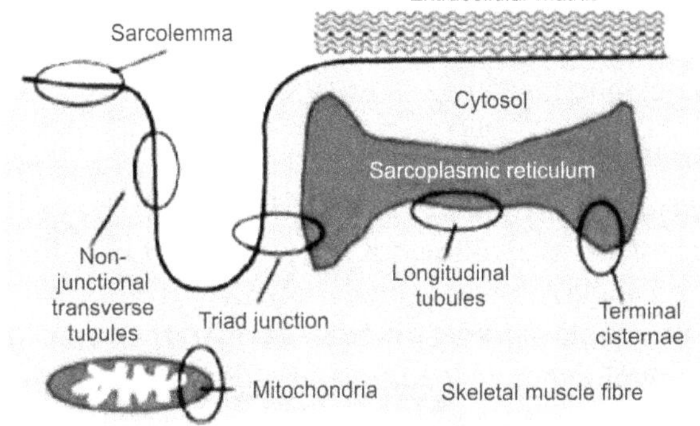

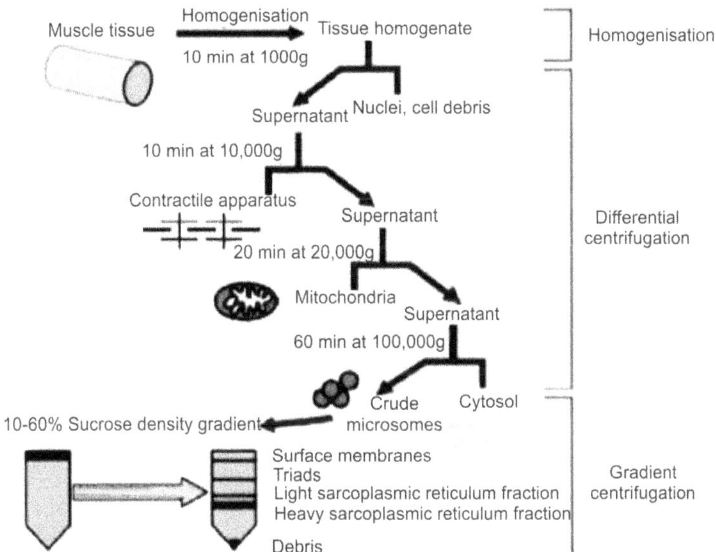

Scheme of the Fractionation of Skeletal Muscle Homogenate into Various Subcellular Fractions. Diagrammatic presentation of the subcellular membrane system from skeletal muscle fibres (a) and a flow chart of the fractionation protocol of these membranes from tissue homogenates using differential centrifugation and density gradient methodology (b).

particles, or even molecules in solution, and causes separations of such matter on the basis ultracentrifugation of differences in weight. Example: Red cells may be separated from plasma of blood, nuclei from mitochondria in cell homogenates, and one protein from another in complex mixtures.

Theodor Svedberg invented the analytical ultracentrifuge in 1923 and won the Nobel Prize in Chemistry in 1926. The vacuum ultracentrifuge was invented by Edward Greydon Pickels which reduced friction generated at high speeds and maintained constant temperature. Spinco introduced the first preparative ultracentrifuge with a centrifugal field of 40,000 **rpm** in 1949. Spinco was later purchased by Beckman instruments in 1954.

5.2.1.1. Preparative Ultracentrifuge

Preparative ultracentrifugation is operated at relative centrifugal fields of up to 9,00,000g. They are available with a wide variety of rotors such as swinging bucket rotor, fixed angle rotor and zonal rotor. Zonal rotors are designed to contain a large volume of sample in a single central cavity rather than in tubes. Some zonal rotors are capable of dynamic loading and unloading of samples while the rotor is spinning at high speed.

The rotor chamber is sealed, evacuated and refrigerated in order to minimize the excessive rotor temperatures generated by frictional resistance between the spinning rotor and the air. Heavy armour plating encapsulates the ultracentrifuge during uncontrolled rotor movements or dangerous vibrations. The modern ultracentrifuge contains temperature regulation system, flexible drive shaft and over-speed control device to prevent unfavourable conditions. A slight rotor imbalance will cause the centrifuge to switch off automatically.

Preparative rotors are used for pelleting of cellular organelles such as mitochondria, microsomes, ribosomes and viruses. They are also be used for sucrose gradient separation of cellular organelles. and caesium salt gradient separation of nucleic acids. After the centrifugation, the rotor is allowed to stop and the gradient is gently pumped out of each tube to isolate the separated components. There is no optical read-out to collect fractions and analyze after each run. The organelles and molecules are separated on the basis of their **sedimentation velocities** or **buoyant densities.**

5.2.1.2. Analytical Ultracentrifuge

Analytical Ultracentrifugation

Analytical Ultracentrifugation (AUC) experiments give us a method for the direct measurement of basic thermodynamic properties of macromolecules in solution. Since sedimentation relies on the principal property of mass and centrifugal force, it is a valuable technique for a wide variety of solution conditions.

Analytical ultracentrifuge can generate a centrifugal field of 2,50,000g in its analytical cell to separate high molecular weight sub-cellular molecules or organelles that differ only slightly in density. It contains a solid rotor which has one analytical cell and one counter balancing cell. The sample is placed in a simple analytical cell at a right angle to the axis of rotation. Then, it is placed in a cylindrical hole on the rotor, balanced by a counter balancing blank cell with the same weight on the opposite side. The rotor is then placed on a spindle attached to a motor below the centrifuge.

Rotors are mostly made out of **titanium alloy** covered with a **polyurethane layer**. The rotor chamber is evacuated with a vacuum pump to reduce air friction, and cooled with a cooling jacket to maintain a constant temperature and prevent overheating. An **optical system** enables the sample to be observed throughout the duration of centrifugation. The light absorption system is called as **Schlieren system**. The beam of light is passed through a quartz window in the bottom of the chamber, through transparent glass windows on the top and bottom of the sample cell, and passes out through another window in the top of the chamber, where it is directed by mirrors to an optical detection device. The light is always ON and a signal is detected only when the sample cell passes through the detection path.

The solution of molecules is then introduced into the sample cell along with a solvent while the rotor is spinning. It requires only small sample size of 20-120 μl at low particle concentrations of 0.01-1g/l. The solvent, **cesium chloride** dissociates into high-density **Cs+** ions that migrate outwards in the direction of the centrifugal force, forming a shallow density gradient. The molecules in solution diffuse centripetally. For eg., when there are two molecular species, the lighter ones migrates faster, and the denser ones lags behind. The discrete change in density at the **interface** between the two regions bends the light passing through the sample cell at that point, and superimposition of adjacent signals is seen as a **spike** at the transition **point**. The size and position of the spike are indications of **quantity** and **molecular weight**, respectively. This is mainly used for **sedimentation equilibrium** analysis. These observations are electronically digitized and stored for further mathematical analysis.

Proteins are separated by ultracentrifugation - very high speed spinning; with possibility of appropriate photography of the protein layers as they form in the centrifugal field, it is possible to determine the molecular weights of proteins. The ultracentrifuge is a centrifuge optimized for spinning a rotor at very high speeds, capable of generating acceleration as high as 19600 km/s^2 (around 50000 rpm).

5.2.2. Kinds of Experiments

The two kinds of experiments commonly performed by ultracentrifuges are sedimentation velocity and sedimentation equilibrium.

5.2.2.1. Sedimentation Velocity

This experiment reports on the **shape** and **molar mass** of the dissolved macromolecules as well as their **size distribution**. Their size resolution depends on the particle radii and rotor speed and ranges from 100 Da to 10 GDa. They are used to study reversible **chemical equilibria** between **macromolecular species**.

5.2.2.2. Sedimentation Equilibrium

These experiments are concerned only with the final steady-state and gives time-independent concentration profile. The sedimentation equilibrium distribution in the centrifugal field is characterized by **Boltzmann distribution.** This experiment is insensitive to the **shape** of the macromolecule, and reports on the **molar mass** of the molecule and **chemical equlibrium** constants of the reaction mixtures.

5.2.3. Application of Ultracentrifuge

The ultracentrifuge is most widely used to

- ☆ Study the proteins, nucleic acids, viruses and other biological macromolecules
- ☆ Study the solution properties of small solutes
- ☆ Measure the molecular weights of solutes
- ☆ Provide the data on molecular weight distributions in polydisperse systems
- ☆ Determine the frictional coefficients and thereby the sizes and shapes of solutes
- ☆ Characterize and separate the macromolecules on the basis of their buoyant densities in density gradients.

Chapter 6

Blotting Technique

Blotting is the technique in which nucleic acids or proteins are immobilized onto a solid support generally nylon or nitrocellulose membranes. Blotting of nucleic acid is the central technique for hybridization studies. Nucleic acid labeling and hybridization on membranes have formed the basis for a range of experimental techniques involving understanding of gene expression, organization, *etc.* Identifying and measuring specific proteins in complex biological mixtures, such as blood, have long been important goals in scientific and diagnostic practice. More recently the identification of abnormal genes in genomic DNA has become increasingly important in clinical research and genetic counseling.

Blotting techniques are used to identify unique proteins and nucleic acid sequences. They have been developed to be highly specific and sensitive and have become important tools in both molecular biology and clinical research.

General Principle

The blotting methods are fairly simple and usually consist of four separate steps:

1. Electrophoretic separation of protein or of nucleic acid fragments in the sample;
2. transfer to and immobilization on paper support;
3. binding of analytical probe to target molecule on paper; and
4. Visualization of bound probe.

Molecules in a sample are first separated by electrophoresis and then transferred on to an easily handled support medium or membrane. This immobilizes the protein or DNA fragments, provides a faithful replica of the original separation, and facilitates subsequent biochemical analysis. After being transferred to the support medium the immobilized protein or nucleic acid fragment is localized by the use of probes, such as antibodies or DNA, that specifically bind to the molecule of interest. Finally, the position of the probe that is bound to the immobilized target molecule is visualized usually by autoradiography.

Blotting is the technique in which nucleic acids or proteins are immobilized onto a solid support generally nylon or nitrocellulose membranes. Nucleic acids or proteins are first separated by gel electrophoresis, transferred to the blotting membranes and probes are used to detect specific molecules from them. There are three types of blotting:

- ✰ Southern blotting
- ✰ Northern blotting
- ✰ Western blotting

6.1. Southern Blotting

Southern blotting is a technique used for the detection of specific DNA fragments in a complex mixture. This technique was invented by Edward Southern, a biologist of Edinburgh University, UK in 1975. The DNA molecules are transferred from an agarose gel onto a membrane and used to locate a specific gene within an entire genome.

6.1.1. Blotting Techniques

There are eight major steps in southern blotting:

1. **Restriction digestion:** The genomic DNA is first digested with restriction endonucleases, which cut high-molecular-weight DNA strands into smaller fragments.
2. **Agarose gel electrophoresis:** The DNA fragments are then separated based on their size by an agarose gel electrophoresis. The gel is then washed with buffer to remove the broken or fragmented residues of agarose and contaminants.
3. **Acid treatment:** If DNA fragments are large in size (>15 kb), it takes longer time

to transfer to	the membrane compared to shorter	DNA fragments
This **depurination** step with an acid (0.25M HCl for 15 min)		takes the purines
out and breaks	the DNA into smaller fragments to allow	efficient transfer
to the membrane. The acid is then neutralized after this step.		

4. **Alkali treatment:** After acid treatment, the gel is placed in 0.25 M NaOH alkali solution to denature the double-stranded DNA into single stranded DNA strands. This improves the binding of the negatively charged DNA

to a positively charged membrane during **hybridization**. It also destroys any residual RNA that are present in the DNA. The gel is then washed with buffer to remove the traces of NaOH.

5. **Blotting:** The denatured DNA in the gel is then placed on the filter paper with wigs dipped in a reservoir containing transfer buffer, Sodium Saline Citrate (SSC). Nitrocellulose or nylon membrane is used for transfer of DNA from the gel. Nitrocellulose has a binding capacity of 100μg/cm, while nylon has a binding capacity of 500 μg/cm. Nylon is less fragile and binds more DNA than nitrocellulose membrane. The membrane is placed on the gel, above which a stack of blotting papers soaked in transfer buffer is placed and gently pressed using glass rod to remove the air that is trapped between the gel and membrane. Then a stack of un soaked blotting papers and a weight of 500 g is placed above them.

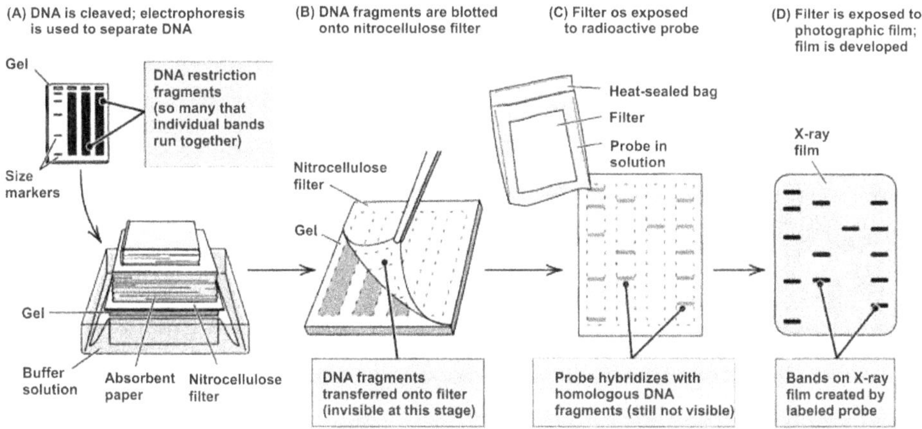

The transfer of DNA from gel to the membrane can take place by three mechanisms.

Capillary system - The transfer buffer is transferred from a region of high water potential to a region of low water potential by capillary action. In this process, the DNA from the gel is transferred onto the membrane and immobilized due to the ion exchange interactions between the negative charge of the DNA and the positive charge of the membrane. This set up is allowed to stand for 12-18 h and the membrane is taken out carefully.

(i) Vacuum blotting system: In vacuum blot apparatus, the vacuum sucks the transfer buffer through the membrane and the gel in a faster speed within an hour.

(ii) Electro transfer system: In this system, an electrical force is used to transfer the DNA from the gel onto the membrane.

6. **Fixation:** There are two widely used methods.
   (i) UVcross linking method: The membrane is exposed to UV rays particularly in case of nylon membrane, so that a **covalent bond** is formed between negatively charged phosphate and thymine residues present in the DNA and the positively charged amino groups on the surface of the nylon membrane.
   (ii) Vacuum oven method: In this method, the nitrocellulose or nylon membrane is baked at 80°C in a vacuum oven or regular oven for 2 h. The use of vacuum oven is a preferrable due to the flammable nature of nitrocellulose membrane.
7. **Hybridization:** Hybridization is the technique in which the nucleic acid immobilized on the membrane is challenged with a known probe. If the probes are made of DNA, they are called as DNA probes. Probes contain single DNA fragment with a specific sequence that are labeled by incorporating radioactive 32 P labeled ATP or tagged with a fluorescent or bioluminescent or chromogenic dye. After the blotting technique is completed, the membrane is exposed to a DNA hybridization probe. If the sequence of DNA probe is complementary to nucleotide sequence on the membrane, it forms hydrogen bond and gets converted into hybrid DNA. This is called base pairing or hybridization. After the hybridization, the membrane is washed using SSC transfer buffer to remove the unbound probes, while bound probes remain attached.
8. **Detection:** Under optimal conditions, 0.1 pg of the DNA can be detected by probing. The regions of hybridization are detected by different methods.
   (i) Autoradiography method: If the probe is radiolabeled 32P or fluorescent labeled, the pattern of hybridization is visualized on X-ray film
   (ii) Biotin streptavidin method: If the probe is labeled by a non-radioactive chromogenic dye, the pattern of hybridization is visualized by development of color on the membrane and measurement by colorimeter.
   (iii) Bioluminescence method: Bioluminescent visualization uses luminesence.

6.1.2. Applications

- ☆ It is used to determine the number of sequences in a genome.
- ☆ It is used to find out specific nucleic acid sequence present in animals.
- ☆ It is used to detect Restriction Fragment Length Polymorphism (RFLP).
- ☆ It is used to detect Variable Number of Tandem Repeat Polymorphism (VNTR), which is the molecular basis of DNA fingerprinting.

6.2. Northern Blotting

Northern blotting is a technique is used for the detection of RNA fragments instead of DNA fragments. The technique is called "Northern" simply because it is similar to "Southern" blotting. Northern blot technique was developed by James Alwine, David Kemp and George Stark of Stanford University, in 1977. There are 3 types of RNA:

1. **tRNA** - transfer RNA, which is active in assembling of polypeptide chains
2. **rRNA** - ribosomal RNA, which is a part of the structure of ribosomes
3. **mRNA** - messenger RNA, which is the product of DNA transcription and used for translation of a gene into a protein

   In these methods, RNA is transferred onto the membrane from the gel. There is no depurination step, as RNA itself is small enough to be transferred easily. The detection of RNA is done with a RNA hybridization probe complementary to part of or the entire target sequence.

6.2.1. Blotting Technique

There are eight steps in Northern blotting:

1. **Isolation of mRNA**: The total RNA is first extracted from a homogenized tissue sample. The mRNA is then isolated through the use of oligo (dT) cellulose chromatography to obtain only those RNAs with a poly(A) tail.
2. **Agarose gel electrophoresis:** The isolated RNA is separated on agarose gels containing formaldehyde as a denaturing agent based on their size. Formaldehyde ensures linear conformation of RNA. Gels are then stained with ethidium bromide (EtBr) and viewed under UV light to observe the quality and quantity of RNA before blotting. Polyacrylamide gel electrophoresis with urea is commonly used for fragmented or microRNAs. An RNA ladder is often run alongside the samples to observe the size of fragments.
3. **Transfer:** The RNA separated on the gel is then transferred to a nylon membrane through a capillary or vacuum blotting system. A nylon membrane with a positive charge is most effective since the negatively charged nucleic acids have a high affinity for them. The transfer saline sodium citrate (SSC) buffer used usually contains formamide or formaldehyde as it lowers the annealing temperature of the probe-RNA interaction and thus prevent RNA degradation at high temperatures.
4. **Fixation:** Once the RNA has been transferred to the membrane, it is immobilized through covalent linkage to the membrane by UV light or heat.
5. **RNA probes:** RNA probes are also called as riboprobes. They are composed of nucleic acids with a complementary sequence to all or part of the RNA of interest. Commonly cDNA is created with labeled primers

for the RNA sequence of interest to act as the probe in the northern blot. Probes can withstand more rigorous washing steps.

6. **Probe labeling:** Probes are labeled either with radioactive isotopes (32P) or with chemiluminescence in which alkaline phosphatase (AP) or horseradish peroxidase (HRP) breakdown chemiluminescent substrates to produce a detectable emission of light. Chemiluminescent labelling occur in two ways, the probe attached to the enzyme or the probe attached with a ligand (biotin) for which the antibody (avidin) is attached to the enzyme. Same membrane can be probed up to five times without a significant loss of the target RNA.
7. **Hybridization:** After a probe has been labeled, it is hybridized to the RNA on the membrane. The membrane is washed to ensure that the probe has bound specifically.
8. **Detection:** The hybrid signals are then detected by X-ray film and can be quantified by densitometry. X-ray film can detect both the radioactive and chemiluminescent signals. Chemiluminescent signals are faster, more sensitive and reduce health hazards associated with radioactive labels.

6.2.2. Applications

- ✰ It is widely used to find out a particular gene's expression pattern between tissues, organs, developmental stages, environmental stress levels, pathogen infection, and over the course of treatment.
- ✰ It is also used to show over expression of oncogenes and down regulation of tumor-suppressor genes in cancerous cells when compared to 'normal'

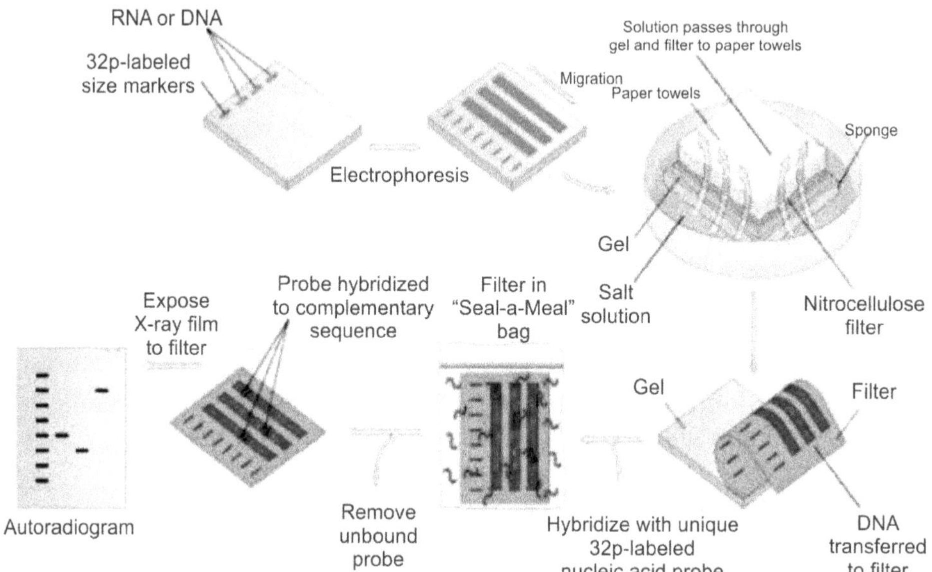

tissue, as well as the gene expression in the rejection of transplanted organs.

6.3. Western Blotting

Western blotting is a method used to transfer the proteins from the polyacrylamide gel onto the membrane and to detect a particular protein of interest using a probe mainly antibiotics. This technique is also referred as "immunoblotting" or "protein blotting". This technique was invented by W. Neal Burnette at the laboratory of George Stark at Stanford. This technique is used in the fields of molecular biology, biochemistry, immunogenetics, *etc.* "Eastern blotting" refers to a technique used for the detection of post-translational modification of protein.

6.3.1. Blotting Technique

There are five major steps in Western blotting:

1. Tissue preparation
2. Gel electrophoresis
3. Transfer
4. Blocking
5. Detection

6.3.1.1. Tissue Preparation

Solid tissues are first broken down mechanically using a blender or homogenizer or by sonication. Buffers are employed to lysis the cells and to solubilize the proteins. Protease and phosphatase inhibitors are often added to prevent the digestion of the sample by its own enzymes. Tissue preparation is done at cold temperatures to avoid protein denaturing. Filtration or centrifugation can be used to separate different cell compartments and organelles.

6.3.1.2. Gel Electrophoresis

Proteins are separated using Sodium dodecyl sulphate-Polyacrylamide (SDS-PAGE) gel electrophoresis based on their molecular weight and electric charge. SDS-PAGE maintains polypeptides in a denatured state. SDS removes the secondary and the tertiary structure of the protein by disrupting the disulfide bonds [S-S] to sulfhydryl groups [SH and SH]. It also covers the denatured proteins with negative charge and moves them to the positively charged electrode through the acrylamide mesh of the gel. Smaller proteins migrate faster through the mesh and the proteins are thus separated according to size, usually measured in kilodaltons, kDa. One lane is usually reserved for a marker or ladder, a commercially available mixture of proteins having defined molecular weights. It is also possible to use atwo-dimensional (2-D) gel which spreads the proteins from a single sample in

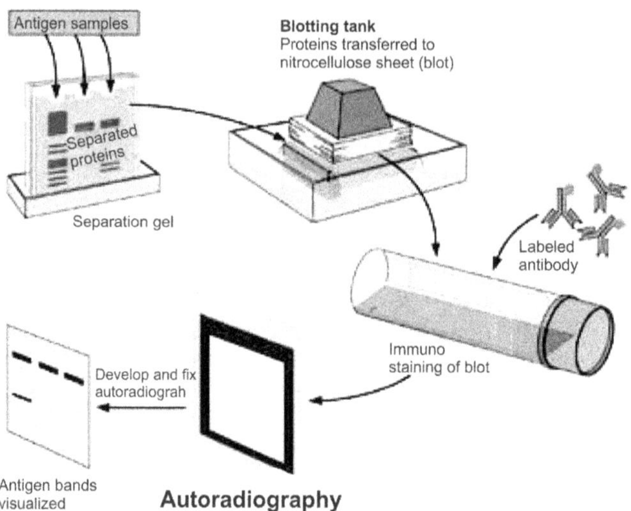

two dimensions. Proteins are separated according to isoelectric point in the first dimension, and according to their molecular weight in the second dimension.

6.3.1.3. Transfer

The separated proteins are transferred from the gel to a membrane made of nitrocellulose or polyvinylidene difluoride(PVDF). The membrane is placed on top of the gel, and a stack of filter papers is placed on top of that. The entire stack is placed in a buffer solution which moves up the paper by capillary action, bringing the proteins with it.

Another method for transferring the proteins is called "electroblotting". It uses an electric current to pull proteins from the gel into the PVDF or nitrocellulose membrane. Protein binding is based upon hydrophobic interactions as well as charged interactions between the membrane and protein. Nitrocellulose membranes are cheaper than PVDF, but are far more fragile and do not stand up for repeated probings. The uniformity and overall effectiveness of transfer of protein from the gel to the membrane can be checked by staining the membrane with Coomassie or Ponceau Sdyes.

6.3.1.4. Blocking

The membrane used for transfer has the ability to bind protein. As both the detection antibodies and the target are proteins, interactions between the membrane and the antibody used for detection of the target protein must be prevented. Blocking of non-specific binding is achieved by placing the membrane in a dilute solution (3-5 per cent) of protein such as bovine serum albumin (BSA) or non-fat dry milk with a minute percentage of detergent such asTween 20 or Triton X-100. This protein attaches to the membrane in all places where the target proteins have not attached. Thus, when the antibody is added, it will not attach other than on the binding sites of the specific target protein.

6.3.1.5. Detection

After the unbound probes are washed away, the blot is ready for detection of the probes that are labeled and bound to the protein of interest. There are several detection methods and the most common detection method is enzyme linked secondary antibody. When the enzyme converts the substrate into an insoluble colored product, it gets precipitated onto the nitrocellulose membrane. The presence of colored band indicates the position of the protein of interest. The protein can be identified by comparison of the blot with the stained gel of the same sample. There are six detection methods.

6.3.1.5.1. Enzyme Linked Secondary Antibody

Two enzymes are commonly used for the detection. They are

- ☆ **Alkaline phosphatase**, which converts colourless 5 bromo 4 chloro indolyl phosphate (BCIP) subtrate into a blue product.
- ☆ **Horse radish peroxidase (HRP)**, which with hydrogen peroxide as a substrate, oxidises either 3 amino 9 ethyl carbazole into an insoluble brown product or 4 chloro-1-napthol into an insoluble blue product.

6.3.1.5.1.2. Chemiluminescence (ECL) Detection

In the presence of hydrogen peroxide and chemiluminescent substrate luminol, HRP oxidizes the luminol with concomitant production of light. The light intensity can be increased 1000 fold with a chemical enhancer. The light emission is then detected by exposing the blot to a photographic film. ECL substrates are also available for use with alkaline phosphatase labeled antibodies.

6.3.1.5.1.3. Radio Labeled Detection

1. 125 I-labeled secondary **antibody** binding to the blot is detected by **autoradiography**
2. Fluorescine-labeled secondary antibody binding to the blot is detected for the fluorescent label by exposing them to the UV light
3. **125I–labeled protein A**. The protein A is purified from Staphylococcus aureus and specifically bound to the Fc region of IgG molecule. This protein binding to the blot is detected by autoradiography

6.3.1.5.1.4. Biotinylated Secondary Antibodies

Biotin is a small molecular weight vitamin that binds strongly to the egg protein avidin. Blot is incubated with biotinylated secondary antibody and then incubated with enzyme conjugated avidin. Multiple biotin molecules can be linked to a single biotinylated antibody molecule and this provides an enhancement of the signal. The enzyme used is usually alkaline phosphatase or horse radish peroxidase.

6.3.1.5.1.5. Quantum Dots

These are engineered semiconductor nanoparticles with diameters of 2-10 nm, which fluoresce when exposed to UV light. Quantum dot nanocrystals comprise a semiconductor core of CdSe (Cadmium selenium) surrounded by a shell of ZnS (Zinc sulpide). This crystal is coated with an organic molecular layer that provides water solubility and conjugation site for biomolecules. Therefore, secondary antibodies will be bound to a quantum dot and the position of binding of the second antibody on the blot is identified by exposing the blot to UV light.

6.3.1.5.1.6. Probes

Probes are also used to identify proteins. Radioactively labeled DNA probes can be used to detect DNA binding proteins on a blot. Blot is first incubated in a solution of radio labeled DNA, then washed to make an autoradiograph of the blot. The presence of radioactive bands is then detected on the autoradiograph and used to identify the positions of the DNA binding proteins on the blot.

6.3.4. Applications

- ☆ It is used to detect anti-HIV antibody in a human serum sample.
- ☆ It is a definitive test for Bovine spongiform encephalopathy (BSE) commonly referred to as 'mad cow disease' as well as Lyme disease.
- ☆ It is used as a confirmatory test for Hepatitis B infection.
- ☆ It can also be applied for mapping restriction sites in single copy gene.

Chapter 7

Molecular Techniques

7.1. Polymerase Chain Reaction (PCR)

7.1.1. Introduction

The **polymerase chain reaction** (**PCR**) is a technology in molecular biology used to amplify a single copy or a few copies of a piece of DNA across several orders of magnitude, generating thousands to millions of copies of a particular DNA sequence. This method is a common technique used in medical and biological research laboratories for a variety of applications such as DNA cloning, DNA-based phylogeny, genetic fingerprinting and diagnosis of diseases. This technique was developed by Kary Mullis, in 1983 and he was awarded the Nobel Prize in Chemistry, in 1993.

Polymerase chain reaction, PCR, is an efficient and cost-effective way to copy or "amplify" small segments of DNA or RNA. Using PCR, millions of copies of a section of DNA are made in just a few hours, yielding enough DNA required for analysis. This innovative yet simple method allows clinicians to diagnose and monitor diseases using a minimal amount of sample, such as blood or tissue.

7.1.2. Technique

Though PCR occurs *in vitro*, or outside of the body in a laboratory, it is based on the natural process of DNA replication. In its simplest form, the reaction occurs when a DNA sample and a DNA polymerase, nucleotides, primers and other reagents (man-made chemical compounds) are added to a sample tube. The reagents facilitate the reaction needed to copy the DNA code.

In addition to detecting diseases in a sample, PCR enables the monitoring of the amount of a virus present, or viral load, in a person's body. In diseases such as hepatitis C or human immunodeficiency virus (HIV) infections, viral load is a good indication of how sick a person may be or how well a person's medicine and treatment is working. Armed with this information, physicians may determine when to start treatment and the person's response to treatment, making treatment personalized to each individual.

There are three clear steps in each PCR cycle, and each cycle approximately doubles the amount of target DNA. This is an exponential reaction so more than one billion copies of the original or "target" DNA are generated in 30 to 40 PCR cycles.

PCR technique is used to amplify a specific region of a DNA strand known as the **target DNA**. It amplifies DNA fragments of up to ~10 kilo base pairs (kb). There are some techniques that can amplify DNA fragments upto 40 kb in size. A heat-stable DNA polymerase known as "**Taq polymerase**" is used for the amplification the target DNA into millions of copies. This enzyme is obtained from the thermophilic bacterium, Thermus aquaticus, which occurs naturally in hot environments of 50 to 80°C. This enzyme can withstand high temperatures of >90°C which is required for separation of the two DNA strands. The enzyme also assembles a new DNA strand by using single-stranded DNA as a **template** and DNA oligonucleotides as the **primers**. Primers are short (oligo) DNA fragments that contain sequences complementary to the target DNA. Primers and dNTPs (deoxyribo nucleotide triphosphates) are required for the synthesis of DNA. Amplification depends on thermal cycling process which consists of cycles of repeated heating and cooling. As the cycle progresses, the DNA generated itself acts as a template for replication, and sets a motion for the chain reaction.

7.1.3. Components of PCR

A basic PCR set up requires the following components and reagents:

- ☆ **DNA template:** This contains the specific DNA target region to be amplified.
- ☆ **Primers:** Two primers that are complementary to the **3'** ends of each of the **sense and anti-sense** strand of the DNA target.
- ☆ **Taq polymerase:** This is a DNA polymerase with a temperature optimum at around 70°C.
- ☆ **Deoxynucleoside triphoshates** (dNTPs): They are the building blocks from which DNA polymerases synthesize a new DNA strand.
- ☆ **Buffer solution:** This provides a suitable chemical environment for optimum activity and stability of the DNA polymerase.
- ☆ **Divalent cations:** Magnesium ions (Mg^{2+}) is generally used. Sometimes manganese ions (Mn^{2+}) are utilized for PCR-mediated DNA mutagenesis.
- ☆ **Monovalent cation:** Potassium ions

7.1.4. PCR Reaction

1. Separating the Target DNA – Denaturation

During the first step of PCR, called denaturation, the tube containing the sample DNA is heated to more than 90 degrees Celsius (194 degrees Fahrenheit), which separates the double-stranded DNA into two separate strands. The high temperature breaks the relatively weak bonds between the nucleotides that form the DNA code.

2. Binding Primers to the DNA Sequence - Annealing

PCR does not copy the all of the DNA in the sample. It copies only a very specific sequence of genetic code, targeted by the PCR primers. For example, Chlamydia has a unique pattern of nucleotides specific to the bacteria. The PCR will copy only the specific DNA sequences that are present in Chlamydia and absent from other bacterial species. To do this, PCR uses primers, man-made oligonucleotides (short pieces of synthetic DNA) that bind, or anneal, only to sequences on either side of the target DNA region.

Two primers are used in step two - one for each of the newly separated single DNA strands. The primers bind to the beginning of the sequence that will be copied, marking off the sequence for step three. During step two, the tube is cooled and primer binding occurs between 40 and 60 degrees Celsius (104 – 140 degrees Fahrenheit).

Step two yields two separate strands of DNA, with sequences marked off by primers.

The two strands are ready to be copied.

3. Making a Copy – Extension

In the third phase of the reaction, called extension, the temperature is increased to approximately 72 degrees Celsius (161.5 degrees Fahrenheit). Beginning at the regions marked by the primers, nucleotides in the solution are added to the annealed primers by the DNA polymerase to create a new strand of DNA complementary to each of the single template strands.

After completing the extension, two identical copies of the original DNA have been made.

The PCR reaction is commonly carried out in a reaction volume of 10–200 μl in small reaction PCR tubes of 0.2–0.5 ml volumes in a **thermal cycler**. The thermal cycler heats and cools the reaction tubes to achieve the temperatures required at each step of the reaction. Most thermal cyclers make use of the **Pletier effect** which permits both heating and cooling of the block, holding the PCR tubes, simply by reversing the electric current. Thin-walled PCR tubes permit favourable thermal conductivity to allow rapid thermal equilibration. Most thermal cyclers have heated lids to prevent condensation at the top of the tube.

PCR Cycles	Target Copies
1	2
2	4
3	8
4	16
5	32
6	64
7	128
8	256
9	512
10	1024
15	32,768
20	1,048,576
25	33,554,432
30	1,073,741,842

After making two copies of the DNA through PCR, the cycle begins again, this time using the new duplicated DNA. Each duplicate creates two new copies and after approximately 30 or 40 PCR cycles, more than one billion copies of the original DNA segment have been made. Because the PCR process is automated, it can be completed in just a few hours.

In a healthcare setting, PCR makes enough copies of target DNA from the clinical sample to allow analysis; the results of these diagnostic and monitoring tests provide clinicians and other healthcare providers with information to guide treatment

1. Real-time PCR
2. Quantitative real time PCR (Q-RT PCR)
3. Reverse Transcriptase PCR (RT-PCR)
4. Multiplex PCR
5. Nested PCR
6. Long-range PCR
7. Single-cell PCR
8. Fast-cycling PCR
9. Methylation-specific PCR (MSP)
10. Hot start PCR
11. High-fidelity PCR
12. In situ PCR
13. Variable Number of Tandem Repeats (VNTR) PCR
14. Asymmetric PCR
15. Repetitive sequence-based PCR
16. Overlap extension PCR
17. Assemble PCR
18. Intersequence-specific PCR(ISSR)
19. Ligation-mediated PCR
20. Methylation –specifin PCR
21. Miniprimer PCR
22. Solid phase PCR
23. Touch down PCR, *etc.*

7.1.5. Steps in PCR

PCR consists of a series of 20-40 repeated temperature changes called "**Cycles**". Most commonly PCR is carried out with cycles that have three temperature steps. Each cycle is often preceded by a single temperature step at a high temperature (>90°C) known as "**Hold**", and another hold at the end, for final product extension or storage known as "**Final hold**". The temperature and the duration in each cycle depend the enzyme used for DNA synthesis, the concentration of divalent ions, dNTPs in the reaction and the melting temperature (Tm) of the primers.

1. **Initial Hold:** This step consists of heating the reaction to a temperature of 94–96 °C for 1 min.
2. **Denaturation:** This step is the first thermal cycling process. It consists of heating the reaction to 94–98 °C for 20–30 sec. In this step, the two strands in a DNA double helix was physically separated into single strands of DNA and this process is known as "**DNA melting**". Separation occurs due to disruption of hydrogen bonds between complementary bases in the double helix.
3. **Annealing:** In this step, the reaction temperature is lowered to 50–65 °C for 20–40 sec.

   This allows annealing of the primers to the single-stranded DNA template through stable hydrogen bonds. The annealing temperature is generally about 3-5°C below the melting point (Tm) of the primers. Taq DNA polymerase then binds to the primer-template hybrid and begins the DNA synthesis.
4. **Extension/elongation:** The temperature at this step depends on the DNA polymerase, particularly the Taq polymerase that has its optimum activity at 72°C. The Taq polymerase synthesizes a new DNA strand complementary to the DNA template strand by the addition of dNTPs that are complementary to the template in 5' to 3' direction by condensing the 5'-phosphate group of the dNTPs with the 3'-hydroxyl group at the end of the extending DNA strand. The duration of extension depends both on the DNA polymerase and the length of the DNA fragment to be amplified. The DNA polymerase can polymerize a thousand bases per minute at its optimum temperature.
5. **Final hold:** The temperature at this step is 4–15 °C for an indefinite time. This is employed for short-term storage of the reaction products.

7.1.6. Detection

The anticipated DNA fragment generated by the PCR is known as **"amplicon".** To check the amplicon, **agarose gel electrophoresis** is performed. The size of the amplicon is determined by comparison with a **DNA ladder,** which contains DNA fragments of known size. DNA ladder has to be run on the gel alongside the PCR products.

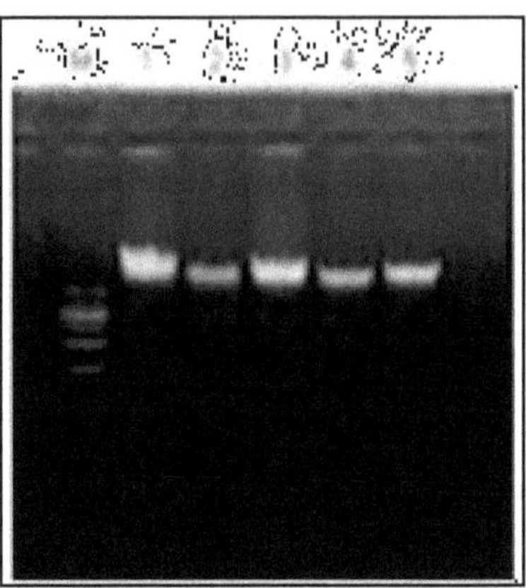

7.1.7. Applications

1. PCR is used in analyzing clinical specimens for the presence of infectious agents, including HIV, hepatitis, malaria, anthrax, *etc.*
2. PCR can provide information on a patient's prognosis, and predict response or resistance to therapy. Many cancers are characterized by small mutations in certain genes, and this is what PCR is employed to identify.
3. PCR is used in the analysis of mutations that occur in many genetic diseases (*e.g.* cystic fibrosis, sickle cell anaemia, phenylketonuria, muscular dystrophy).
4. PCR is also used in forensics laboratories and is especially useful because only a tiny amount of original DNA is required, for example, sufficient DNA can be obtained from a droplet of blood or a single hair.
5. PCR is an essential technique in cloning procedure which allows generation of large amounts of pure DNA from tiny amount of template strand and further study of a particular gene.
6. The Human Genome Project (HGP) for determining the sequence of the 3 billion base pairs in the human genome, relied heavily on PCR.
7. PCR has been used to identify and to explore relationships among species in the field of evolutionary biology. In anthropology, it is also used to understand the ancient human migration patterns. In archaeology, it has been used to spot the ancient human race. PCR commonly used by Paleontologists to amplify DNA from extinct species or cryopreserved fossils of millions years and thus can be further studied to elucidate on.

7.2. Cloning

7.2.1. Introduction

Cloning is a process used to create an exact genetic replica of another cell, tissue, or organism. The copied material that has the same genetic material as the original is referred to as a clone. The term clone is derived the Greek word, which means "trunk or branch". In biology, **cloning** is the process of producing similar populations of genetically identical individuals that occurs in nature when organisms such as bacteria, insects or plants reproduce asexually.

Cloning in biotechnology refers to processes used to create copies of DNAfragments (molecular cloning), cells (cell cloning), or organisms. The term **clone**, invented by J. B. S. Haldane, is derived from the Ancient Greek word "twig", referring to the process whereby a new plant can be created from a twig. In horticulture, the spelling clon was used until the twentieth century; the final e came into use to indicate the vowel is a "long o" instead of a "short o".

Natural cloning is a natural form of reproduction that has allowed life forms to spread for more than 50 thousand years. It is the reproduction method used by plants, fungi, and bacteria, and is also the way that clonal colonies reproduce themselves.

7.2.2. History

The modern cloning techniques involving nuclear transfer have been successfully performed on several species. Notable experiments include:

- ☆ Tadpole: (1952) Robert Briggs and Thomas J. King had successfully cloned northern leopard frogs: thirty-five complete embryos and twenty-seven tadpoles from one-hundred and four successful nuclear transfers.
- ☆ Carp: (1963) In China, embryologist Tong Dizhou produced the world's first cloned fish by inserting the DNA from a cell of a male carp into an egg from a female carp. He published the findings in a Chinese science journal.
- ☆ Mice: (1986) A mouse was successfully cloned from an early embryonic cell. Soviet scientists Chaylakhyan, Veprencev, Sviridova, and Nikitin had the mouse "Masha" cloned.
- ☆ Sheep: Marked the first mammal being cloned (1984) from early embryonic cells by Steen Willadsen. Megan and Morag cloned from differentiated embryonic cells in June 1995 and Dolly the sheep from a somatic cell in 1996.
- ☆ Rhesus monkey: Tetra (January 2000) from embryo splitting
- ☆ Pig: the first cloned pigs (March 2000). By 2014, BGI in China was producing 500 cloned pigs a year to test new medicines.
- ☆ Gaur: (2001) was the first endangered species cloned.
- ☆ Cattle: Alpha and Beta (males, 2001) and (2005) Brazil
- ☆ Cat: **CopyCat** "CC" (female, late 2001), Little Nicky, 2004, was the first cat cloned for commercial reasons
- ☆ Rat: Ralph, the first cloned rat (2003)

- ✫ Mule: Idaho Gem, a john mule born 4 May 2003, was the first horse-family clone.
- ✫ Horse: Prometea, a Haflinger female born 28 May 2003, was the first horse clone.
- ✫ Dog: Snappy, a male Afghan hound was the first cloned dog (2005).
- ✫ Wolf: Snuwolf and Snuwolffy, the first two cloned female wolves (2005).
- ✫ Water buffalo: Samrupa was the first cloned water buffalo. It was born on 6 February 2009, at India's Karnal National Diary Research Institute but died five days later due to lung infection.
- ✫ Pyrenean ibex (2009) was the first extinct animal to be cloned back to life; the clone lived for seven minutes before dying of lung defects.
- ✫ Camel: (2009) Injaz, is the first cloned camel.
- ✫ Pashmina goat: (2012) Noori, is the first cloned pashmina goat. Scientists at the faculty of veterinary sciences and animal husbandry of Sher-e-Kashmir University of Agricultural Sciences and Technology of Kashmir successfully cloned the first Pashmina goat (Noori) using the advanced reproductive techniques under the leadership of Riaz Ahmad Shah.

Ethical Issues of Cloning

There are a variety of ethical positions regarding the possibilities of cloning, especially human cloning. While many of these views are religious in origin, the questions raised by cloning are faced by secular perspectives as well. Perspectives on human cloning are theoretical, as human therapeutic and reproductive cloning are not commercially used; animals are currently cloned in laboratories and in livestock production.

Advocates support development of therapeutic cloning in order to generate tissues and whole organs to treat patients who otherwise cannot obtain transplants, to avoid the need for immunosuppressive drugs, and to stave off the effects of aging. Advocates for reproductive cloning believe that parents who cannot otherwise procreate should have access to the technology.

Opponents of cloning have concerns that technology is not yet developed enough to be safe and that it could be prone to abuse (leading to the generation of humans from whom organs and tissues would be harvested), as well as concerns about how cloned individuals could integrate with families and with society at large.

Religious groups are divided, with some opposing the technology as usurping "God's place" and, to the extent embryos are used, destroying a human life; others support therapeutic cloning's potential life-saving benefits.

Cloning of animals is opposed by animal-groups due to the number of cloned animals that suffer from malformations before they die, and while food from cloned animals has been approved by the US FDA, its use is opposed by groups concerned about food safety.

7.2.3. Types of Cloning

There are three different types of cloning:

- Molecular cloning
- Cellular cloning
- Organism cloning

7.2.3.1. Molecular Cloning

Molecular cloning or DNA cloning is a process of creating multiple copies of a defined DNA sequence. It amplifies the DNA fragments containing whole genes as well as any DNA sequence such as promoters, non-coding sequences and randomly fragmented DNA. It is used in a wide array of applications starting from **genetic fingerprinting** to large scale **protein production**.

Molecular cloning is also known as recombinant DNA technology. In this process, a piece of DNA or a gene from one organism is transferred to a self replicating genetic element such as a bacterial plasmid, which is a cloning vector. The cloning vector is a small piece of DNA into which a foreign DNA fragments can be inserted. The plasmid with inserted DNA sequence is then transferred to a bacterium, where multiple copies of the same piece of DNA or gene are generated. The sequence must be linked to the **origin of replication** of the bacterial plasmid for amplification. The origin of replication is a sequence of DNA capable of directing the propagation any linked sequence. It is used in a wide array of biological experiments and practical applications ranging from genetic fingerprinting to large scale protein production.

Cloning of any DNA fragment essentially involves four steps:

1. Fragmentation

Fragemenation means breaking apart a strand of DNA. The DNA fragment of suitable size has to be isolated, amplified into their product and cut into smaller fragments using restriction enzymes.

2. Ligation

Ligation means gluing together pieces of DNA in a desired sequence. In this step, the DNA fragment is inserted into a plasmid vector. The vector must be first linearised using same restriction enzymes used to cut the DNA and incubated with the DNA fragment under appropriate conditions with an enzyme called **DNA ligase**.

3. Transfection

Transfection means inserting the newly formed pieces of DNA into cells. A number of techniques are available for transfection and they include chemical sensitivation of cells, electroporation, optical injection and biolistics.

4. Screening/Selection

The transfected cells are then cultured in the desired medium for selecting the cells that are successfully transfected with the vector construct containing the desired DNA sequence. Modern cloning vectors contain selectable antibiotic resistance markers, which allow only cells in which the vector has been transfected to grow. In addition, the cloning vectors also contain colour selection markers, which provide blue/white screening on **X-gal** medium. Recombinant plasmid will appear white or colourless in the presence of X-gal, whereas an intact non-recombinant plasmid will be blue since its gene is fully functional and not disrupted. Further investigation can be accomplished by means of PCR, restriction fragment analysis and/or DNA sequencing.

7.2.3.2. Cellular Cloning

Cloning Unicellular Organisms

Cloning a cell means to derive a population of cells from a single cell. In the case of unicellular organisms such as bacteria and yeast, this process is remarkably simple and essentially only requires the inoculation of the appropriate medium. However, in the case of cell cultures from multi-cellular organisms, cell cloning is an arduous task as these cells will not readily grow in standard media.

A useful tissue culture technique used to clone distinct lineages of cell lines involves the use of cloning rings (cylinders). According to this technique, a single-cell suspension of cells that have been exposed to a mutagenic agent or drug used to drive selection is plated at high dilution to create isolated colonies, each arising from a single and potentially clonal distinct cell. At an early growth stage when colonies consist of only a few cells, sterile polystyrene rings (cloning rings), which have been dipped in grease, are placed over an individual colony and a small amount of trypsin is added. Cloned cells are collected from inside the ring and transferred to a new vessel for further growth.

Cloning Stem Cells

Somatic-cell nuclear transfer, known as SCNT, can also be used to create embryos for research or therapeutic purposes. The most likely purpose for this is to produce embryos for use in stem cell research. This process is also called "research cloning" or "therapeutic cloning." The goal is not to create cloned human beings (called "reproductive cloning"), but rather to harvest stem cells that can be used to study human development and to potentially treat disease. While a clonal human blastocyst has been created, stem cell lines are yet to be isolated from a clonal source.

Therapeutic cloning is achieved by creating embryonic stem cells in the hopes of treating diseases such as diabetes and Alzheimer's. The process begins by removing the nucleus (containing the DNA) from an egg cell and inserting a nucleus from the adult cell to be cloned. In the case of someone with Alzheimer's disease, the nucleus from a skin cell of that patient is placed into an empty egg.

The reprogrammed cell begins to develop into an embryo because the egg reacts with the transferred nucleus. The embryo will become genetically identical to the patient. The embryo will then form a blastocyst which has the potential to form/ become any cell in the body.

The reason why SCNT is used for cloning is because somatic cells can be easily acquired and cultured in the lab. This process can either add or delete specific genomes of farm animals. A key point to remember is that cloning is achieved when the oocyte maintains its normal functions and instead of using sperm and egg genomes to replicate, the oocyte is inserted into the donor's somatic cell nucleus. The oocyte will react on the somatic cell nucleus, the same way it would on sperm cells.

The process of cloning a particular farm animal using SCNT is relatively the same for all animals. The first step is to collect the somatic cells from the animal that will be cloned. The somatic cells could be used immediately or stored in the laboratory for later use. The hardest part of SCNT is removing maternal DNA from an oocyte at metaphase II. Once this has been done, the somatic nucleus can be inserted into an egg cytoplasm. This creates a one-cell embryo. The grouped somatic cell and egg cytoplasm are then introduced to an electrical current. This energy will hopefully allow the cloned embryo to begin development. The successfully developed embryos are then placed in surrogate recipients, such as a cow or sheep in the case of farm animals.

SCNT is seen as a good method for producing agriculture animals for food consumption. It successfully cloned sheep, cattle, goats, and pigs. Another benefit is SCNT is seen as a solution to clone endangered species that are on the verge of going extinct. However, stresses placed on both the egg cell and the introduced nucleus are enormous, leading to a high loss in resulting cells.

Cloning a cell means to derive a population of cells from a single cell. Tissue culture technique is used to clone distinct lineages of cell lines. In this technique, a single-cell suspension is exposed to a mutagenic agent or drug to drive selection. These cells are plated at high dilution to create isolated colonies. Each colony arises from a single and clonally distinct cell. At the early growth stage when the colonies consist of only a few of cells, sterile polystrene rings or cloning rings dipped in grease are placed over an individual colony and a small amount of trypsin is added. The cloned cells are then collected from inside the ring and transferred to a new vessel for further growth.

7.2.3.3. Organism Cloning

Organism cloning (also called reproductive cloning) refers to the procedure of creating a new multicellular organism, genetically identical to another. In essence this form of cloning is an asexual method of reproduction, where fertilization or inter-gamete contact does not take place. Asexual reproduction is a naturally occurring phenomenon in many species, including most plants (see vegetative reproduction) and some insects. Scientists have made some major achievements with cloning, including the asexual reproduction of sheep and cows. There is a lot of ethical debate over whether or not cloning should be used. However, cloning, or asexual propagation, has been common practice in the horticultural world for hundreds of years. There are three different types of "organism cloning".

1. Embryo cloning
2. Reproductive cloning
3. Therapeutic cloning

7.2.3.3.1. Embryo Cloning

Embryo cloning is known as "embryo splitting". This technique is similar to the natural process of creating identical twins. Twins occur in nature just after fertilization of an egg cell by a sperm cell. When the fertilized egg called a zygote tries to divide into a two-celled embryo, the two cells sometimes separate and each cell continues to divide on its own and develops into a separate individual. As the two cells come from the same zygote, the resulting individuals are genetically identical.

In artificial embryo "twinning" or "splitting", this process occurs in a petri dish instead of mother's body. It involves manual separation of individual cells from an early embryo. Sperm and an egg cell are mixed together on a glass petri dish. After conception, the zygote is allowed to develop into a blastula stage, which contains a hollow mass of cells. In-vitro fertilization (IVF) is performed at this blastula stage to increase the number of available embryos. A chemical is added to the petri dish to remove the "zona pellucida" covering. This chemical also provides nutrients to the cells to promote cell division. The blastula is then divided into individual cells which are then deposited on individual dishes. The cells are then coated with an artificial zona pellucida and allowed to divide and develop into embryos. Interruption of the zygote at the two cell stage gives best results. The resulting embryos are then implanted into a surrogate mother and allowed to develop into an individual. As all the embryos come from the same zygote, the resulting individuals are genetically identical monozygotic twins.

7.2.3.3.2. Reproductive Cloning

Reproductive cloning is used for the creation of an animal that has the same DNA as another animal. It has been successfully used to clone Dolly, the sheep. In this type of cloning, the DNA from an ovum is removed and replaced with the DNA from a somatic cell removed from an adult animal. The fertilized ovum or pre-embryo, is then implanted in a uterus and allowed to develop into a new animal. This technique is known as **Somatic Cell Nuclear Transfer (SCNT)**. In these techniques, the resulting offspring will be genetically identical to the donor and not the surrogate, unless the donated nucleus is taken from a somatic cell of the surrogate.

There are two variations in this technique:

- Roslin technique
- Honolulu technique

SCNT refers to the transfer of the nucleus from a somatic cell to an egg cell. A somatic cell is any cell in the body other than the two types of reproductive germ cells *i.e.* sperm and egg. Somatic cell could be a blood cell, heart cell, skin cell, *etc.* In mammals, every somatic cell has two complete sets of chromosomes, whereas the reproductive germ cells have only one complete set of chromosomes. The nucleus in a cell is an enclosed compartment that contains all information in the form of DNA that each cell needs to form an organism.

In SCNT process, the egg from a female donor is first taken and its nucleus is removed to create an **enucleated egg**. A somatic cell which contains DNA is taken from the person to be cloned. Then, the enucleated egg is fused together with the somatic cell. The embryo thus created begins to divide normally. It is then implanted into the uterus of the surrogate mother. The clone produced will be an exact genetic replica of the adult that donated the somatic cell nucleus to the egg. The difference between fertilization and SCNT lies in origin of two sets of chromosomes. In fertilization, the sperm and egg both contain one set of chromosomes; and the resulting zygote ends up with two sets - one from the father (sperm) and one from the mother (egg). In SCNT, the egg cell's single set of chromosomes is removed and replaced by the nucleus from a somatic cell, which already contains two complete sets of chromosomes.

7.2.3.3.2.1. Roslin Technique

Roslin technique is developed by researchers of Roslin Institute, who created the sheep, Dolly. In this process, somatic cells with nuclei intact are allowed to grow and divide. They are then deprived of nutrients to induce the cells into a suspended or dormant stage. The enucleated egg cell is then placed in close proximity to a somatic cell and both cells are shocked with an electrical pulse. The developing embryo is then implanted into a surrogate and allowed to develop.

7.2.3.3.2.2. Honolulu Technique

Honolulu technique is developed by Dr. Teruhiko Wakayama at the University of Hawaii. In this method, the nucleus from a somatic cell is removed and injected into the enucleated egg cell. The egg is then given a bath in a chemical solution. The developing embryo is then implanted into a surrogate and allowed to develop.

7.2.3.3.2.3. Problems in Reproductive Cloning

The clones are sometimes not strictly identical since the somatic cells may contain mutations in their nuclear DNA. Also, the mitochondrial DNA from the donor egg interferes and the mitochondrial genome will not be the same as that of the nucleus donor cell in the clone. This leads to cross-species nuclear transfer and results in nuclear-mitochondrial incompatibilities.

7.2.3.3.3. Therapeutic Cloning

- ☆ Therapeutic cloning is identical to the reproductive cloing in the initial stages of cloning. In this technique, the stem cells are removed from the pre-embryo to produce tissue or a whole organ for transplant back into the person who supplied the DNA and the pre-embryo dies during the process. The first successful therapeutic cloning was accomplished by Advanced Cell Technology, a Biotech company in Worcester, MA, USA in 2000. This cloning is also known as "Research cloning" because the embryos produced are mainly used in stem cell research to study human development and to treat disease.
- ☆ In therapeutic cloning, the SCNT process is employed. The egg cell from a female donor is first taken and its nucleus is removed to create an **enucleated egg**. An egg cell consists of all basics required for creating a pre-embryo. The somatic cell is taken from the sick person. The enucleated egg is then fused together with the somatic cell by giving an electrical shock. The resultant embryo is allowed to grow for 14 days to produce many stem cells. The stem cells are then removed from the embryo and encouraged to grow into a piece of human tissue or a complete human organ for transplant. This tissue or organ would be transplanted into the patient. Stem cells can be used to develop into replacement organs such as heart, liver, pancreas, skin, *etc.*

7.2.3.3.3.1. Hurdles in Therapeutic Cloning

Therapeutic cloning has not yet been accomplished in the laboratory or clinic. There were four main hurdles to overcome.

- ☆ Stem cells have to be "successfully isolated and grown in the laboratory."
- ☆ They have to be encouraged to "turn into specific cell types."
- ☆ They have to be proven usable in treating patients with diseases, injuries, or disorders.
- ☆ The transplanted tissue must develop normally and must not represent significant "risks to the patient."

7.2.4. Advantages and Disadvantages of Cloning

Advantages of Cloning

1. Solution to infertility
2. Provides organs for transplantation
3. Provides treatments for variety diseases
4. Genetic modification/engineering
5. Healthiness of infants
6. Better understanding of genetic diseases

7. Helps improve lives
8. Protects Endangered Species
9. Improves food supply

Disadvantages of Cloning

1. Element of Uncertainty
2. Inheriting diseases
3. Potential for Abuse

7.3. Cell Culture

7.3.1. Introduction

Cell culture refers to the culturing of cells derived from multicellular eukaryotes, especially animal cells under controlled conditions. In 1885, Wilhelm Roux maintained a portion of the medullary plate of an embryonic chicken in a warm saline solution for several days and established the principle of tissue culture. Later, the cell culture techniques advanced significantly in the 1940s and 1950s mainly to support research in virology. Viruses grown in cell cultures are purified and used mainly for the manufacture of vaccines. Salk polio vaccine was one of the first vaccine mass-produced using cell culture techniques by John Franklin Enders, Thomas Huckle Weller, and Frederick Chapman Robbins.

Cell culture refers to the removal of cells from an animal or plant and their subsequent growth in a favorible artificial environment. The cells may be removed from the tissue directly and disaggregated by enzymatic or mechanical means before cultivation, or they may be derived from a cell line or cell strain that has already been established. Primary culture Primary culture refers to the stage of the culture after the cells are isolated from the tissue and proliferated under the appropriate conditions until they occupy all of the available substrate (*i.e.*, reach confluence). At this stage, the cells have to be subcultured (*i.e.*, passaged) by transferring them to a new vessel with fresh growth medium to provide more room for continued growth. After the first subculture, the primary culture becomes known as a cell line. Cell lines derived from primary cultures have a limited life span (*i.e.*, they are finite), and as they are passaged, cells with the highest growth capacity predominate, resulting in a degree of genotypic and phenotypic uniformity in the population. If a subpopulation of a cell line is positively selected from the culture by cloning or some other method, this cell line becomes a cell strain. A cell strain often acquires additional genetic changes subsequent to the initiation of the parent line. Normal cells usually divide only a limited number of times before losing their ability to proliferate, which is a genetically determined event known as senescence; these cell lines are known as finite. However, some cell lines become immortal through a process called transformation, which can occur spontaneously or can be chemically or virally induced. When a finite cell line undergoes transformation and acquires the ability to divide indefinitely, it becomes a continuous cell line. Culture conditions vary widely for each cell type, but the artifical environment in which the cells are cultured invariably consists of a suitable vessel containing a substrate or medium that supplies the essential nutrients (amino acids, carbohydrates, vitamins, minerals), growth factors, hormones, and gases (O_2, CO_2), and regulates the physicochemical milieu (pH, osmotic pressure, temperature). Most cells are anchorage dependent and must be cultured while attached to a solid or semi-solid substrate (adherent or monolayer culture), while others can be grown floating in the culture medium (suspension culture). If a surplus of cells are available from sub culturing, they should be treated with the appropriate protective agent (*e.g.*, DMSO or glycerol) and stored at temperatures below –130°C (cryopreservation) until they are needed. Morphology of cells in culture Cells in culture can be divided in to three basic categories based on their shape and appearance (*i.e.*, morphology).

1. Fibroblastic (or fibroblast-like) cells are bipolar or multipolar, have elongated shapes, and grow attached to a substrate.
2. Epithelial-like cells are polygonal in shape with more regular dimensions, and grow attached to a substrate in discrete patches.
3. Lymphoblast-like cells are spherical in shape and usually grown in suspension without attaching to a surface

7.3.2. Types of Cultures

Cells can be grown as suspension or adherent cultures.

Suspension Culture

Some cells especially blood cells naturally live in suspension. Cultivation of these cells is done by suspending the cells in the medium. Suspension culture is also done for certain cells that do not express their characteristics in the adherent form. These cell lines are modified to survive in suspension cultures to be grown to a higher density. Absence of serum components in the growth medium helps to prevent adhesion.

Adherent Culture

Adherent cells require a surface for their growth. Tissue culture plastic or microcarrier, which are coated with extracellular matrix components are commonly used to increase adhesion properties and to provide other signals needed for growth and differentiation. Most cells derived from solid tissues are adherent. Another type of adherent culture is known as "**organotypic culture**" which involves growing cells in a three-dimensional environment.

7.3.3. Steps in Cell Culture

The specific requirements of a cell culture laboratory depend mainly on the type of research conducted; all cell culture laboratories have the common requirement of being free from pathogenic microorganisms (*i.e.*, asepsis), and share

Adherent Culture	*Suspension Culture*
Appropriate for most cell types, including primary cultures.	Appropriate for cells adapted to suspension culture and a few other cell lines that are nonadhesive (*e.g.*, hematopoietic).
Requires periodic passaging, but allows easy visual inspection under inverted microscope.	Easier to passage, but requires daily cell counts and viability determination to follow growth patterns; culture can be diluted to stimulate
Cells are dissociated enzymatically (*e.g.*, TrypLE™ Express, trypsin) or mechanically.	Does not require enzymatic or mechanical dissocation
Growth is limited by surface area, which may limit product yields.	Growth is limited by concentration of cells in the medium, which allows easy scale-up.
Requires tissue-culture treated vessel.	Can be maintained in culture vessels that are not tissue-culture treated, but requires agitation (*i.e.*, shaking or stirring) for adequate gas exhange.
Used for cytology, harvesting products continuously, and many research applications.	Used for bulk production, batch harvesting

some of the same basic equipment that is essential for culturing cells. This section lists the equipment and supplies common to most cell culture laboratories, as well as beneficial equipment that allow the work to be performed more efficiently or accurately, or permits wider range of assays and analyses. Note that this list is not all inclusive; the requirements for any cell culture laboratory depend on the type of work conducted.

Basic Equipment

- ✰ Cell culture hood (*i.e.*, laminar-flow hood or biosafety cabinet)
- ✰ Incubator (humid CO_2 incubator recommended)
- ✰ Water bath
- ✰ Centrifuge
- ✰ Refrigerator and freezer (–20°C)
- ✰ Cell counter
- ✰ Inverted microscope
- ✰ Liquid nitrogen (N_2) freezer or cryostorage container
- ✰ Sterilizer (*i.e.*, autoclave) Expanded equipment
- ✰ Aspiration pump (peristaltic or vacuum)
- ✰ pH meter
- ✰ Roller racks (for scaling up monolayer cultures)
- ✰ Confocal microscope
- ✰ Flow cytometer
- ✰ EG bioreactors
- ✰ Cell cubes Additional supplies

- Cell culture vessels (*e.g.*, flasks, Petri dishes, roller bottles, multiwell plates)
- Pipettes and pipettors
- Syringes and needles
- Waste containers
- Media, sera, and reagents
- Cells

The major requirement of a cell culture laboratory is the need to maintain an aseptic work area that is restricted to cell culture work. Although a separate tissue culture room is preferred, a designated cell culture area within a larger laboratory can still be used fort sterile handling, incubation, and storage of cell cultures, reagents, and media. The simplest and most economical way to provide aseptic conditions is to use a cell culture hood (*i.e.*, biosafety cabinet). The cell culture hood provides an aseptic work area while allowing the containment of infectious splashes or aerosols generated by many microbiological procedures. Three kinds of cell culture hoods, designated as Class I, II and III, have been developed to meet varying research and clinical needs.

Class I cell culture hoods offer significant levels of protection to laboratory personnel and to the environment when used with good microbiological techniques, but they do not provide cultures protection from contamination. They are similar in design and air flow characteristics to chemical fume hoods.

Class II cell culture hoods are designed for work involving BSL-1, 2, and 3 materials, and they also provide an aseptic environment necessary for cell culture experiments. A Class II biosafety cabinet should be used for handling potentially hazardous materials (*e.g.*, primate-derived cultures, virally infected cultures, radioisotopes, carcinogenic or toxic reagents).

Class III biosafety cabinets are gas-tight, and they provide the highest attainable level of protection to personnel and the environment. A Class III biosafety cabinet is required for work involving known human pathogens and other BSL-4 materials. Air-flow characteristics of cell culture hoods Cell culture hoods protect the working enviroment from dust and other air born contaminants by maintaining a constant, unidirectional flow of HEPA-filtered air over the work area. The flow can be horizontal, blowing parallel to the work surface, or it can be vertical, blowing from the top of the cabinet onto the work surface. Depending on its design, a horizontal flow hood provides protection to the culture (if the air flowing towards the user) or to the user (if the air is drawn in through the front of the cabinet by negative air pressure inside). Vertical flow hoods, on the other hand, provide significant protection to the user and the cell culture. Clean benches Horizontal laminar flow or vertical laminar flow "clean benches" are not biosafety cabinets; these pieces of equipment discharge HEPA-filtered air from the back of the cabinet across the work surface toward the user, and they may expose the user to potentially hazardous materials. These devices only provide product protection. Clean benches can be used for certain clean activities, such as the dust-free assembly of sterile equipment or electronic devices, and they should never be used when handling cell culture materials or drug formulations, or when manipulating potentially infectious materials.

Cell culture hood layout A cell culture hood should be large enough to be used by one person at a time, be easily cleanable inside and outside, have adequate lighting, and be comfortable to use without requiring awkward positions. Keep the work space in the cell culture hood clean and uncluttered, and keep everything in direct line of sight. Disinfect each item placed in the cell culture hood by spraying them with 70 per cent ethanol and wiping clean. The arrangement of items within the cell culture hood usually adheres to the following right-handed convention, which can be modified to include additional items used in specific applications.

- ✰ A wide, clear work space in the center with your cell culture vessels
- ✰ Pipettor in the front right and glass pipettes in the left, where they can be reached easily
- ✰ Reagents and media in the rear right to allow easy pipetting
- ✰ Small container in the rear middle to hold liquid waste.

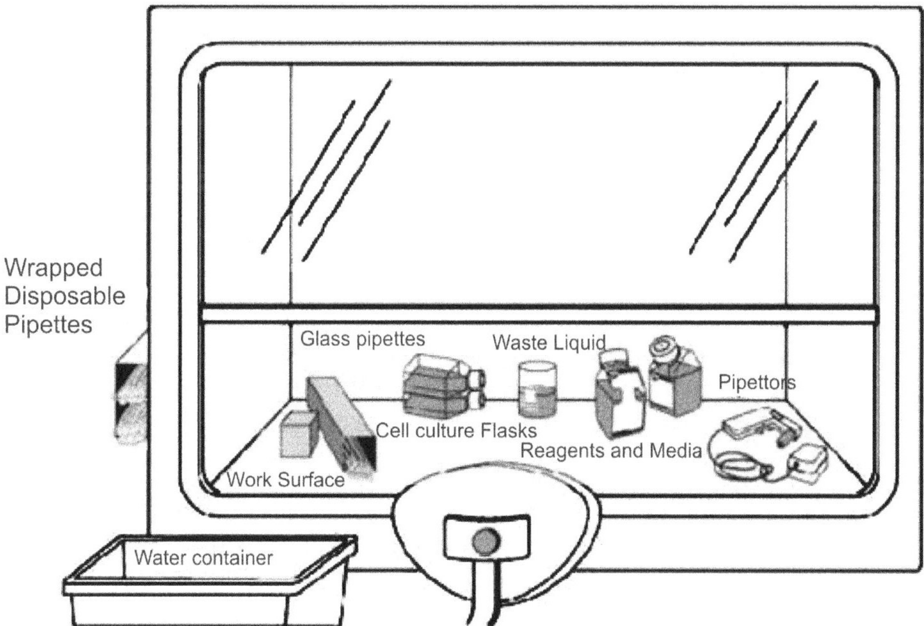

The steps involved in cell culture includes

1. Isolation of cells

2. Maintenance of cells
3. Manipulation of cultured cells

7.3.3.1. Isolation of Cells

Cells are isolated from tissues in several ways. Blood cells can be easily purified but only the white cells can be grown by cell culture.

Mononuclear cells can be released from soft tissues by enzymatic digestion with the help of enzymes such as collagenase, trypsin or pronase. Soft tissue pieces can be placed in growth media, and the cells that grow out are available for culture. This method is known as "**explant culture**".

Cells that are cultured directly from a tissue are known as "**primary cells**". Most primary cell cultures have limited lifespan. After a certain number of population doublings, these cells undergo the process of senescence and stop dividing but retain the viability.

Random mutation or deliberate modification such as artificial expression of the telomerase gene can help an immortalised cell line to acquire the ability to proliferate indefinitely.

7.3.3.2. Maintenance of Cells

Certain factors are responsible for the maintenance of cells. They are:

1. **Temperature and CO_2:** Cells have to be maintained at an appropriate temperature (37°C) and gas mixture (5 per cent CO_2) in a cell incubator. Culture conditions vary widely for each cell type.
2. **Growth medium:** The most important factor for the maintenance of cells is the growth medium. Growth media with different pH, glucose concentration, growth factors, and nutrients are available. Growth factors are often derived from animal blood, such as calf serum. Antibiotics such as penicillin and streptomycin and antifungals such as Amphotericin B can be added to the growth media to prevent contamination.
3. **Contamination:** Contamination of the cell culture with viruses or prions is the main problem because of the blood-derived ingredients in the growth media. To minimize or eliminate this problem, it is better to use the animal blood from countries with minimum BSE/TSE risk such as Australia and New Zealand or to use of purified nutrient concentrates derived from serum in place of whole animal serum for cell culture.

7.3.3.3. Manipulation of Cultured Cells

Manipulation of cultured cells becomes important as the cells that grow generate several issues:

- ✰ Nutrient depletion in the growth media
- ✰ Accumulation of apoptotic/necrotic (dead) cells.

- ☆ Cell-to-cell contact that stimulate cell cycle arrest.
- ☆ Cell-to-cell contact that stimulate cellular differentiation.

The common manipulations that are carried out in cell culture are change of media, passaging and transfection of cells. Manipulations are carried out in a biosafety hood or laminar flow cabinet to exclude contamination by micro-organisms.

a. Change of Media

As the cells grow, they undergo metabolic processes. Acid is produced which in turn decreases the pH. Addition of a pH indicator to the medium helps to identify the nutrient depletion. Then, the media can be removed directly by aspiration and replaced with the new media.

b. Passaging

Passaging is also known as "**subculture**" or "**splitting cells**". This involves transfer of a small number of cells into a new vessel. Cells that are subcultured regularly can be grown for a longer time, as it avoids the senescence associated with prolonged high cell density. Suspension cultures are easily passaged with a small amount of culture containing a few cells diluted in a larger volume of fresh media. In adherent cultures, cells are first detached with a mixture of trypsin-EDTA and then a small number of detached cells are grown as a new culture.

c. Transfection and Transduction

Introduction of foreign DNA by transfection in another common method for cell manipulation. This helps the cells to express a protein of interest. The transfection of RNAi constructs is used to suppress the expression of a particular gene/protein. Foreign DNA can also be inserted into cells using viruses by the process called "**transduction**".

7.3.4. Applications

- ☆ Tissue culture and engineering: Cell culture is essential for tissue culture and tissue engineering. The major application of human cell culture is in stem cell industry where mesenchymal stem cells are cultured and cryopreserved for future use.
- ☆ Vaccines: Cell cultures are used for the production of vaccines for polio, measles, mumps, rubella and chicken pox. Recombinant DNA-based vaccines, such as human adenovirus are some of the recent developments.
- ☆ Viral culture methods: Culture of viruses require the culture of cells of mammalian, plant, fungal or bacterial origin as hosts. Whole wild type viruses, recombinant viruses or viral products may be generated in cell types other than their natural hosts under the right conditions.

☆ It is also used in drug screening and development, and large scale manufacturing of biological compounds (*e.g.*, vaccines, therapeutic proteins).

7.4. Hybridoma Technology

7.4.1. Introduction

Hybridoma technology is a technology of forming hybrid cell lines (called hybridomas) by fusing an antibody-producing B cell with a myeloma (B cell cancer) cell that is selected for its ability to grow in tissue culture and for an absence of antibody chain synthesis. The antibodies produced by the hybridoma are all of a single specificity and are therefore monoclonal antibodies (in contrast to polyclonal antibodies). The production of monoclonal antibodies was invented by Cesar Milstein and Georges J. F. Kohler in 1975. They shared the Nobel Prize of 1984 for Medicine and Physiology with Niels Kaj Jerne, who made other contributions to immunology. The term hybridoma was coined by Leonard Herzenberg during his sabbatical in Cesar Milstein's laboratory in 1976/1977.

Group of antibodies having different specificity produced from different clone of B cell is referred as polyclonal antibodies. Antigens with multi-epitopes injected into mice. After an incubation period serum is isolated from mice. It will contain antibodies against different epitopes and they are also produced from different clone of cells. Serums with antibodies are known as antiserum which is produced against an injected antigen like antitoxin which is nothing but antibody produced against toxins.

Antigens with multivalent epitope injected into mice. After incubation period B cells are isolated. They are hybridized with myeloma cells with the help of fusogens different types of B cells are separated. Antibodies from each type act as mono clonal antibodies. Monoclonal antibodies are produced mainly by hybridoma technology.

7.4.2. Technique

It refers to the production of hybridoma cells by a technology. In this technology, cells of interest hybridized with cancer cells. The aim of this technology is to provide immortality to the cells of interest. When cells of interest fused with cancer cells with the help of fusogens like sendaivirus or polyethylene glycol (PEG), first cytoplasmic membrane of two cells are fused to form heterokaryon *i.e.* cells with two different nucleuses. Then the two nucleuses combined to form cybrid or hybrid cells. Hybrid cells contain combination of both cell nuclear materials but maintaining species usual chromosomal number by loss of extra chromosome. In cybrid cells genetic material of either of the cell maintained completely and the other one is lost completely *i.e.* cybrid cells are hybrids considering the cytoplasmic component.

Of the three possibility of cells like unfused cells, hybridoma cells and cybrid cells, hybridoma cells selected by using HAT (Hypoxanthine Aminopterin Thymidine) medium.

HAT Medium Selection

HAT selection depends on the fact that mammalian cells can synthesize nucleotides by two different pathways: the de novo and the salvage pathways. The de novo pathway in which a methyl or formyl group is transferred from an activated from of tetrahydrofolate, is blocked by Aminopterin, a folic acid analog. When the de novo pathway is blocked, cells utilize the salvage pathway, which bypasses the aminopterin block by converting purines and pyrimidines directly into DNA. The enzymes catalyzing the salvage pathway include hypoxanthine-guanine phosphorribosyl transferase (HGPRT) and thymidine kinase (TK). A mutation in either of these two enzymes blocks the salvage pathway. HAT medium contains Aminopterin to block the de novo pathway and hypoxanthine and thymidine to allow growth via the salvage pathway. When two types of cells, one with a mutation in TK and the other with a mutation in HGPRT are fused, only the hybrid cells will contain the full complement of necessary enzymes for growth on HAT medium via the salvage pathway. Thus only hybrid cells will grow in HAT medium.

Monoclonal antibodies produced in the following steps:

1. Preparation of Cells
2. Fusion of cells
3. Selection of Hybridoma cells
4. Culturing of selected hybridoma cells
5. Screening of culture for antibody production
6. Propagation of screened colony
7. Isolation of antibodies

1. Preparation of Cells

For the production of monoclonal antibodies, cells should be properly prepared. B cell and myeloma cells are used for monoclonal antibody production. Genetic machinery of B cells should be modified as Ig+, HGPRT---- and TK+. This is achieved through site directed mutagenesis process. These B cells are actually isolated from mice to which antigen of our interest is injected and it also posses the ability of producing antibody against antigen of our interest. Genetic machinery of myeloma cells are modified as Ig----, HGPRT+ and TK----. The reason for genetic modification is to provide antibody producing ability to B cells and for selecting hybrid cell from the culture after hybridization. Myeloma cells are used to provide immortality property.

2. Fusion of Cells

After preparing B cell and myeloma cell, they are fused with the help of sendaivirus or polyethylene glycol (PEG). These fusogens fuse the cells to produce fused B cells-myeloma cells, cybrid and hybrids. Some unfused cells may also present in the medium.

3. Selection of Hybrid Cells

After fusing the cells with the help of fusogens they are placed in a culture medium with selection medium of HAT. Aminopterin inhibits de novo nucleotide biosynthesis. When de novo synthesis inhibited cells can use salvage pathway to produce nucleotides for that they require HGPRT and TK. Since one of this enzyme is absent in B cells, myeloma cells and cybrids, they are unable to survive in the selection medium. Those hybrid cells which are having functional HGPRT and TK are found to be capable of growing in selection medium. It can be otherwise called as that they are selected in selection.

4. Culturing of Selected Hybridoma Cells

After hybridoma cells are selected, they are cultured in microtiter wells in such a way that each well consists of single hybridoma cells. So that, mono clone is produced *i.e.* Group of cells derived from single hybridoma cells.

5. Screening Cells for Antibody Production

Eventhough hybridoma cells survived in the selection medium and they also form clones, there ability to form antibody is not tested yet. This property is tested using ELISA or RIA tests. Those clones which have the ability to produce antibody of our interest are allowed for next step of monoclonal antibody production.

6. Propagation of Screened Clones

It is nothing but culturing or allowing selected clone cells to grow in invitro or invivo conditions respectively. In invitro technique clone cells are cultured in tissue culture flask with suitable medium. The production rate is found to be 10 - 100micrograms per ml. In invivo method, selected clones are injected into the peritoneal cavity of histocompatible mice and allowed to multiply and produce antibody. The rate of antibody production is 25 milligram per ml.

7. Isolation of Antibodies

In either of above methods, samples are collected and from these monoclonal antibodies are separated by affinity chromatography and they are used for different studies.

7.4.3. Types of Monoclonal Antibodies

A. Immunotoxins

B. Chimeric Immunotoxins

C. Humanized antibodies

D. CDR grafted antibodies

E. Heterconjugate antibodies

A. Immunotoxins

It is nothing but the complex of antibodies and toxins like ricin, diphtheria toxin

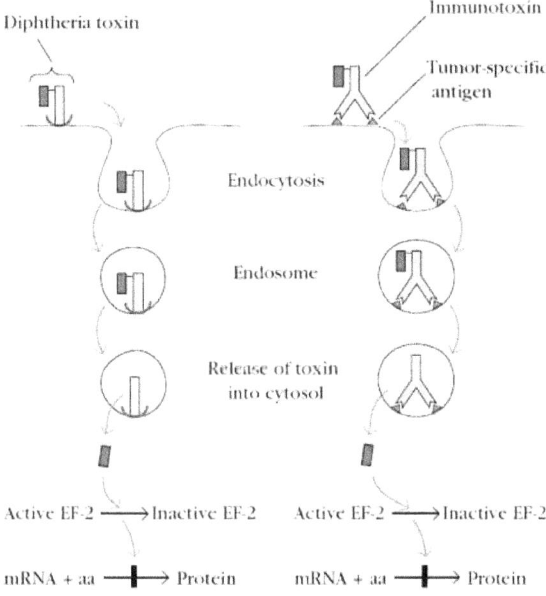

etc., i.e. toxins linked to immunoglobins at their Fc region. Immunotoxins are mainly used against tumors. When Immunotoxins for a particular tumor cells injected, they reached the target cell with the help of antibodies. After their attachment to tumor cells, it is taken up by tumor cells through endocytosis. After this, cellular metabolism interfered with the help of toxins and cells died. Because of the use of monoclonal antibodies, toxins are directed towards specific tumor cells.

B. Chimeric Immunotoxins

The term chimera used because unlike Immunotoxins, toxin replaces one of the constant region domain of Ig especially carboxy terminal domain of H chain. It can also be used to treat cancers. It is advantageous than Immunotoxins in the sense that it won't activate complements and consequent inflammatory reaction after binding to the antigens. This is because of the lack of Fc region in chimeric Immunotoxins.

C. Humanized Antibodies

When antibodies from mice injected as anti-tumor agents, they can stimulate immunogenic reactions against them in the host. This can be avoided to certain extent using humanized antibodies. Humanized antibodies contain constant region of both L and H chain derived from humans and variable regions derived from mice. It is named so because of the presence of human gene coding regions. It can be used as diagnostic tool also.

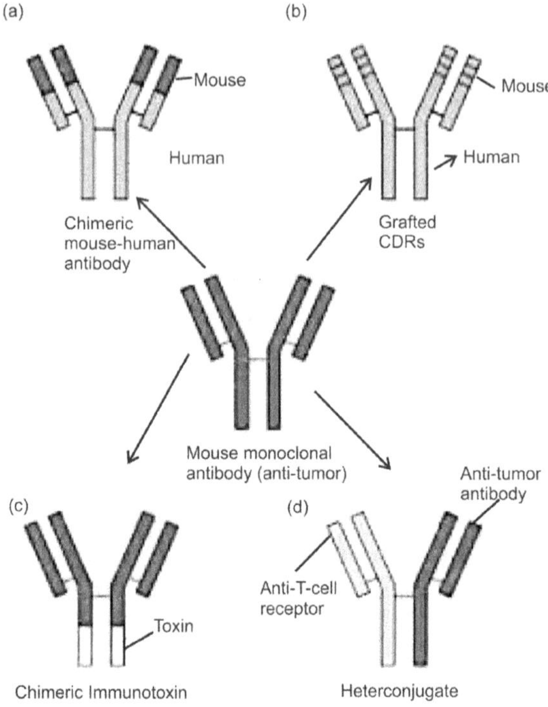

D. CDR Grafted Antibodies

It is the modified form of humanized antibodies *i.e.* hypervariable region of both light and heavy chains variable region gene only derived from mice. The remaining genes derived from human. It is used against cancer and it also act as immunosuppressor. Usually in such cases, antibodies are directed against CD3 or CD4 components. Because once these components blocked, activation of cell mediated immunity affected.

E. Heterconjugate Antibodies

The word hetero means that antibodies with different specificity *i.e.* different Paratopes having different specificity in single antibody. This type of antibody specifically used to treat cancer. In this case, one paratope of antibody directed against tumor cells whereas other paratope is directed against cytotoxic cells usually Tc cells. These antibodies bring effective cytotoxic cells close to tumor cells, so that they are easily lysed by cytotoxic mechanisms.

7.4.4. Applications

Monoclonal antibodies are proving to be very useful as diagnostic, imaging, and therapeutic reagents in clinical medicine. Mainly monoclonal antibodies applied in two fields namely,

I. Diagnostic Field
II. Therapeutic Field

I. Diagnostic Field

a. Leucocyte Identification
b. Lymphocyte subset determination
c. HLA typing
d. Viral detection and sub typing
e. Parasitic determination
f. Polypeptide hormone detection
g. Detection of cancer with tumor marker determination
h. Detection of cardiac myosin in cardiac injury
i. Pregnancy detection

In the above mentioned conditions, the basic principle is production of antibodies against antigen and identification or quantification of antigen and antibody complex.

II. Therapeutic Field

a. Anti-tumor therapy
b. Immunosuppression
c. Fertility control
d. Drug toxicity reversal

a. Anti-tumor Therapy

In anti-tumor therapy, antibodies against tumor antigen produced and they are converted into either Immunotoxins, or chimeric Immunotoxins or Heterconjugate antibodies. When these antibodies utilized, they damage tumor cells and tumor growth controlled. These anti-tumor antibodies are called as "magic bullets".

b. Immunosuppression

During transplantation between partially incompatible individuals, host versus graft rejections are suppressed using monoclonal antibodies against TCR, BCR, Co-receptor complex and cytokines *etc.*, Hypersensitivity reactions are also treated with blocking monoclonal antibodies.

c. Fertility Control

By producing antibodies against HCG or trophoblast, fertility controlled.

d. Drug Toxicity Reversal

Toxicity produced by drugs is treated using monoclonal antibodies against drugs, so that the functions of drugs are blocked and effect reversed.

References

Lakshmanan P.T., Suseela Mathew, Anandan R., Asha K.K. and Niladri Sekhar Chatterjee, Biochemical Analysis Of Seafood, CIFT publication.

Niessen, W.M.A. 2007. Liquid chromatography – Mass Spectrometry, 3rd Edn., Boca Raton F.L.:CRC press.

Pyell, U. 2006. Electro kinetic chromatography: Theory, Instrumentation and Applications, John Wiley, New York.

Simonian, M.H. and Smith, J. A. 2006. Spectrophotometric and colorimetric determination of protein concentration. Current Protocols in Molecular Biology. Chap. 10, Unit 10. 1A, Wiley Inter science, New York.

Wilson, K. and Walker, J. 2010. Principles and Technique of Biochemistry and Molecular Biology. 7th Edn., Cambridge University Press, New Delhi. p. 744.

www.ingramcontent.com/pod-product-compliance
Ingram Content Group UK Ltd.
Pitfield, Milton Keynes, MK11 3LW, UK
UKHW021953270726
14060UKWH00002B/488

9 789388 173209